FOUNDATIONS

Design & Technology

Resistant Materials Technology

KS3

Paul Anderson
Jeff Draisey

Nelson Thornes

Published in 2011 by:
Nelson Thornes Ltd
Delta Place
27 Bath Road
CHELTENHAM
GL53 7TH
United Kingdom

11 12 13 14 15 / 10 9 8 7 6 5 4 3 2 1

A catalogue record for this book is available from the British Library

ISBN 978 1 4085 0812 1

Cover photograph: BananaStock/Punchstock
Page make-up by Pantek Arts Ltd, Maidstone
Printed and bound in Poland by Drukarnia Dimograf

Acknowledgements

Alamy: pv (MBI), p31T (Image Broker), p74TL (Acnoncc), p76T (Jurgen Hanel), p76B (Nikcreates), p77T (Elizabeth Whiting and Associates), p77B (Duncan Snow), p79B (Aleksander Ugorenkov), p81T (Shinypix), p81B (Acnoncc); **Alyson Blanchard:** p32B; **AP Valves:** p39 – all; **Dyson:** p23T, p23 – bottom five images, p33 – all; **Fotolia:** p9, p13, p17 – all, p19B, p42T, p42B, p47M, p47R, p48B, p49, p51M, p54T, p55T, p62T, p62M, p80T; **Getty Images:** p21, p37T, p37B (Digital Vision), p46 (Photo Alto), p51B, p56B (Barry Willis), p66 (Thinkstock), p72T, p73L, p73M (Bloomberg); **Innova Systems:** p23M; **iStockphoto:** pvi, p3L, p3R, p12M, p12BL, p18T, p18B, p36T, p36B, p43TL, p43TR, p43BL, p43BR, p44L, p44R, p47L, p48T, p53B, p54UM, p54LM, p55LM, p55B, p56T, p60, p61, p62B, p64, p65T, p65B, p70, p72B, p73R, p74TR, p80M, p80B; **Jeff Draisey:** p74B, p75; **Mark Boulton:** p32T, p69T, p69B; **Paul Anderson:** p40, p41; **Photolibrary:** p19T; **Roger Smith:** p31B; **Ruth Amos:** p78; **Science Photo Library:** p6 and p50T (Maximilian Stock Ltd), p51T (Eye of Science), p52 (Pascal Goetcheluck), p53T (Chris Priest and Mark Clarke); **Series P:** p71.

Thank you to all the students of Hazel Grove High School for their examples of Resistant Materials Technology used in this book.

Contents

Introduction to Resistant Materials

Resistant materials

Our world is full of products that are designed and made from resistant materials. From the moment you wake in the morning, your whole day will be affected by resistant materials. From the bed that you climb out of, to the drawers that your clothes are in, to the kettle that you use to make a drink, the car that takes you to school, the table that you are sitting at, the plate that your dinner is served on ... Every one of these products and many, many more are designed and made from resistant materials.

The structure of this book

This book is divided into two sections. The first section (Chapters 1–6) takes you through the process of designing a product. This will help you to build up the skills that you need to design successful products. The second section (Chapters 7 and 8) will help you to improve your knowledge of materials and making activities. This will support you to successfully make products. Chapter 9 provides some case studies on famous designers, designing for others and product development.

Designing

You need to understand the different design needs that a product must meet. This will help you to generate ideas for effective products. You will learn techniques that will help you to become a creative designer. These will also help you to present your ideas to others. You will also learn how to prepare for the making of products using resistant materials.

A designer must consider how their product might affect others. This can range from how many people are employed to make the product to how people are affected by its use. Obtaining the resources needed to make the products will also have an effect on the environment. By reading this book, you will become not only a better, more creative designer, but also more aware of how the products you use affect our society and the environment that we live in.

Case study

How mobile phones affect society

Mobile phones have had a huge impact on our society. In the UK alone, over 150,000 people are employed to make, sell or support the use of them.

Mobile phones have made it easier for people to stay in contact with each other, for social reasons or for work. They can be of great help in times of emergency. For example, an ambulance can be called from the scene of an accident, allowing it to get to any injured person faster; or if a car breaks down in a remote area, it is much easier to call for help.

However, mobile phones can also cause problems. Some people become angry due to the noise from other people using them in public places or on public transport. There has been an increase in phone-related crime – over 700,000 phones are stolen each year. Accidents have been caused by people using mobile phones when driving cars.

Making

To turn designs into working products you will need to know about the materials that could be used to make them. You will need to understand the properties of the materials, their advantages and disadvantages. You will need to know about how different processes can be safely used to make finished products. You will also learn how products are made in industry.

....and finally

Resistant Materials is an exciting and very rewarding subject. It will involve you in a great deal of decision making and practical work. You will need to plan ahead and become very organised. In the end, you should finish up with a wide range of knowledge, skills and understanding. This will be very useful to you over the coming years and, hopefully, you will have designed and made some products that you can be really proud of.

1.1   Why develop new products?

Objectives

- Understand what is meant by 'design process'.
- Be able to identify a need that a product is designed to meet.

Key terms

Design process: a sequence of activities carried out to develop a product.

Need: what the product you are designing must do.

Want: features that you would like the product to have.

The design process

When you are designing a new product, there are a number of things that need to be done to create it. You must decide what is needed, how to make it and, after making it, test it to see that it meets the needs of the user. One way of showing these activities is as a series of tasks. This is referred to as the **design process** (Figure **A**).

In practice, the design process is often not carried out in order. As you go through it, you find out more about the final product. This can change what was needed at an earlier step. For example, you might find out that you don't have the materials that you wanted to use, so you would have to change your design. Sometimes you might have to jump back to earlier steps several times during the process.

Identifying the need

The first step in the design process is to identify the **need**. The need is what the product you are designing must do. One way to do this is to write down a 'problem'. If you turn this into a positive statement, it becomes the need (Figure **B**). The need should be quite general, to give the designer the opportunity to come up with a creative or unusual solution.

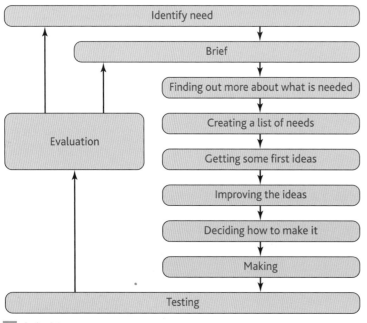

A   A design process

It is important to separate needs from wants. A **want** is something that you would *like* the product to have, but the problem could still be solved without it. For example, a need could be for 'a personal transport system'. A want would be 'it should be blue because I like that colour'. The problem could still be solved if the solution was pink, green or purple.

Needs and designing products

A single product might have to meet several different needs. For example, a toy for a small child might have to be educational, safe and low cost. Sometimes meeting one need makes it harder to achieve one of the other needs. For example, we might need the toy to be very strong. However, we also need it to be low cost, and this might stop us using the strongest material. When needs contradict each other, the designer will have to make a decision about which need is more important.

Another important point is that needs can change. This might be due to changes in society or developments in technology. For example, in recent years, it has become a need to use recycled materials for many products, to reduce damage to the environment.

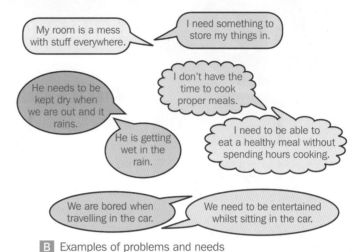

B Examples of problems and needs

Case study

Mobile phones

One solution to the need for people to be able to keep in touch when they are out and about is a mobile phone.

The first mobile phone service started in Japan in 1978. These phones were often bigger and heavier than a house brick. Once the original need had been met, people developed more needs – for smaller, lighter and less expensive phones. These needs have been met because of improvements in batteries, electronics and the machines used to make them.

Mobile phones now include a wide range of other features, such as alarms, cameras, MP3 players and games. They are available in a wide variety of styles and colours, meeting a wide range of customer 'wants' as well as the need.

Can you imagine how our needs and wants might change again in the future, and how mobile phones might develop in response to this?

C Early mobile phone (left) and modern mobile phone (right)

Activities

1 Make a list of items that you have used today – for example, the clothes you are wearing, furniture and electrical equipment. For each item, identify the needs that it was designed to meet.

2 A teenager, a person with a visual impairment (blind), and an old age pensioner can have different needs for the same activity. For one of the following activities, explain how their needs to carry it out might be different:

- using a mobile phone
- preparing a meal
- travelling on public transport.

Summary

● The design process is the series of activities that are carried out to create a product.

● The first step in the design process is to identify the need.

Understanding the design brief

What is a design brief?

After a need has been identified, the next activity is to write the **design brief**. The design brief is a short statement of what needs to be designed.

The design brief is given to the person who will design the product. It is often only one paragraph long. A good brief will state:

- what the product must do
- who will use it
- who might buy it, known as the **market** for the product
- the things that might limit the design, known as **constraints**
- some important features of the product.

The example below is a design brief for a child's toy.

'educational toy':
This is the need. It is the function that the product must carry out.

'cost should be similar to other toys':
This is another constraint. The designer may have some great ideas that would cost too much to make. They would have to reject these ideas and choose one that could be made at a lower cost.

'parents and relatives':
This is the market for the product – the people who might buy it. Often, this is the same as the users for the product, but not always.

> There is a need for an **educational toy** for children aged three to **five years old**. It would be bought by **parents and relatives**. The cost **should be similar to other toys** available in the shops. It should be suitable to be made as **a one-off product**. It must be **safe to use**. It should be made using **sustainable material**.

'children aged three to five years':
This is the user of the product.

'safe to use':
This is a good example of an important product feature.

'made using sustainable materials':
This is another important product feature. Sustainable material means natural materials that can be replaced without damaging the environment. For example, this might be wood or cotton.

'suitable to be made as a one-off product':
This is a constraint because it limits what the designer can do. They will have to design the toy so that it can be made with tools that can make just one product at a time, rather than big industrial machines.

A A design brief for a child's toy

Analysing the design brief

The next task is to analyse the design brief. The aim is to identify all the information that you need to know to be able to design the product. This involves asking lots of questions about what is needed, for example:

- What does the product have to do? How might it do this?
- Who will use it? Who else might be affected by it?
- How will it be used?
- Where will it be used?
- What could it be made from? How could it be made?

You should explore each answer in as much detail as possible. Normally, you won't know all the answers at this stage. However, this analysis tells you what you need to find out in the next step of the design process.

One way to analyse the brief is to create a 'word web' or spider diagram. This starts with the need in the centre. You then make branches for each question about what is needed. Many of these will have further branches off them, with more questions about the detailed needs of the design.

Activities

1 Choose a product that you are familiar with. For example, a chair, a tin-opener or a television. Write the design brief that you think was used for this product.

2 Below is a design brief for a product. Analyse the brief using a word web.

Office workers require easy access to their mobile phones. Design a mobile phone 'cradle' that will sit on the office desk. It will be made in large quantities and sold at a very low cost. It should be made from recycled materials. It should have a theme based on a favourite sports team.

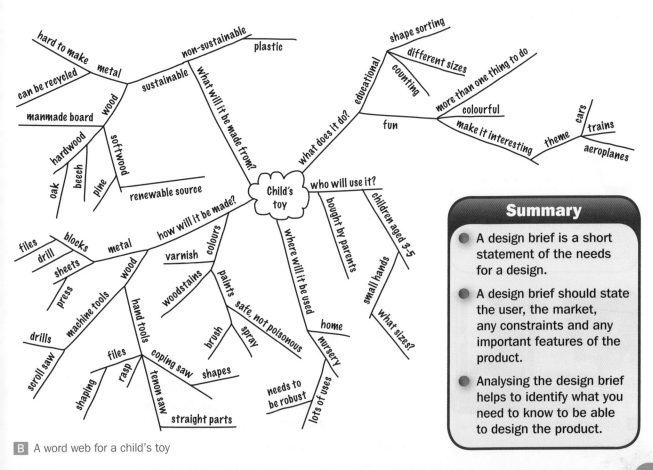

B A word web for a child's toy

Summary

- A design brief is a short statement of the needs for a design.

- A design brief should state the user, the market, any constraints and any important features of the product.

- Analysing the design brief helps to identify what you need to know to be able to design the product.

- Understand the purpose of carrying out research.
- Be able to identify a range of different research activities.

Key terms

Research: collecting the information you need to be able to design the product.

Primary research: finding out the information you need by yourself.

Secondary research: finding out the information you need by using data that someone else has put together.

Questionnaire: a list of questions used to find out what lots of users want from the product.

Link

See **1.2 Understanding the design brief** for more information on analysing the design brief.

A   Testing the strength of a material

Why do research?

The purpose of **research** is to collect all the information that you need to be able to design the product. You will probably have identified what you need to know when you analysed the design brief.

The research you carry out should be relevant and analysed:

- **Relevant** means that you should only investigate the things that you need to know. For example, there would be no point investigating the properties of metals if you have to make your product out of plastic.
- **Analysed** means that you should draw conclusions from your research. This means explaining what that piece of research means for your design. For example, each piece of research could be finished with a statement that starts: 'As a result of this, I have decided that my product …'

Types of research

There are two types of research: primary and secondary. **Primary research** is where you find out the information needed yourself. **Secondary research** is where you use information that someone else has put together.

For example, if you are investigating how you might make your product:

- primary research might include testing different manufacturing processes to see what they can do
- secondary research might include using books or watching videos to find out about the processes.

Primary research normally takes more time than secondary research. It can be much quicker to read about something than to carry out tests to find out the information. However, primary research can be focused exactly on the product and information that you are investigating. Secondary research is sometimes more general, although it might still be able to tell you the information that you need.

Primary research

Some examples of primary research are:

- using questionnaires to ask users what they want from your product
- testing different materials to find out their properties
- taking existing products apart to find out how they are made.

Secondary research

Some examples of secondary research are:

- looking up material properties in a textbook
- visiting exhibitions and museums
- interviewing people with expert knowledge. For example, you might ask someone who has already made a similar product why they used certain design features or processes.

Questionnaires

A **questionnaire** is primary research. It is simply a list of questions. Ideally, as many users as possible should answer the questionnaire to make sure that the design will be suitable for lots of people.

Questionnaires should start by asking if the person answering is a potential user of the product. They might ask about age, gender and whether they would use the type of product being designed. They should then ask a series of questions to find out about each piece of information needed. For example, which colour they like, how much they might pay for the product, etc.

It is useful to give the person answering a choice of possible answers for each question. This means that it is easy to show the answers in bar graphs or pie charts, which will help you to analyse the results.

Questionnaire

Hi, my name is Mike Lloyd and I am going to design and manufacture a portable chessboard that's appealing to the teenage eye. I would be very grateful if you answer a few of my questions.

1. How old are you? (Please tick)
   13-14 ☐ 15-16 ☐ 17-18 ☐ 19+ ☐

2. What gender are you?
   Male ☐ Female ☐

3. How often do you play chess?
   Every day ☐ Weekly ☐ Monthly ☐ Hardly ever ☐

4. What size would you want the chessboard?
   Palm size ☐ Lap size ☐ Desktop size ☐

5. The chess board will be able to compact itself. How would you like it to compact?
   Fold down middle ☐ In a box ☐ Roll up into tube ☐
   Draws for the pieces ☐ Inflatable ☐ Other ....

6. Which material(s) would you like it to be made out of?
   Wood ☐ Glass ☐ Metal ☐ Plastic ☐ Rubber ☐

B Example of a questionnaire

1.4 Product analysis

When you are designing a new product, there will probably already be products that meet similar needs. You can use these products as a source of information to help you design your product.

Product analysis involves investigating existing products. However, it is not just about describing them. It is about understanding why they are designed in a certain way.

- If you identify the good features of the product, you might be able to use these in your design.
- If you identify ways that the product could be improved, this might help you to create an even better product.

ACCESS FM

ACCESS FM is a way of remembering what you should investigate when analysing a product. Each letter stands for a different thing you should analyse: **aesthetics**, customer, cost, environment, size, safety, **function**, and materials and manufacture.

Aesthetics	• What colour is it? • What shape is it? • What texture does the surface have? Is it smooth or rough? Are any edges sharp or rounded? • Does the product look attractive? Why?
Customer	• Who might use the product? • Why would they use this rather than another similar product?
Cost	• How much does it cost to buy? • Is it good value?
Environment	• Is it environmentally friendly? • Is it made from recycled materials? • Can it be recycled?
Size	• What size is it? • How long, wide and tall is it? (Hint: measure it and write down the sizes.) • Why is it this size? • If the size was made bigger or smaller, would it work better?
Safety	• How has the user been protected from harm when using it? • Are there any sharp edges, small parts that could be swallowed, loose wires, etc? • Does the product meet recognised safety standards and regulations? How do you know?
Function	• What will the product be used for? • How would it be tested to make sure it is suitable? • How could it be improved to make it work better? • Where will it be used? • Are there any maintenance activities needed? These are things you have to do to keep it working well, like changing batteries, cleaning it or replacing broken parts
Materials and manufacture	• What materials is it made from? • Would different materials work (or look) better? • How was it made? • Why was it made this way? • If it has different parts, how are these joined together?

Case study

Toy train

This is an example of a product analysis for a child's toy train:

Aesthetics	The train uses lots of bright colours, so that children will find it interesting to look at. It has rounded edges and smooth surfaces because they look good and make it feel nicer when you hold it.
Customer	It would be used by small children aged less than 3 years. It would be bought by their parents.
Cost	The cost of this train is about £10, so that it is about the same cost as other toys for children of this age.
Environment	The train is made from wood – it cannot be recycled but it is a natural material which can be replaced by growing new trees.
Size	The train and blocks are that size so that they fit in a child's hand. The holes in the blocks need to be bigger than a finger so that the child's finger cannot get stuck in them.
Safety	It has no sharp edges, so that children cannot cut themselves on it. All the parts that can be taken off are quite big, so that no one can choke on them.
Function	The train has a motor and two carriages. The carriages have lots of blocks of different shapes and colours that can be taken off and put back on in different ways. It will help children to learn about shapes and how they fit together. The surface is coated in coloured or clear paint so that it is easy to clean. This makes it last longer and it's easier to remove germs.
Materials and manufacture	The train is made from wood. The different parts have been attached together by making holes that they can slot into and using glue, because this gives a much stronger joint that just gluing. It was probably made using computer-controlled machines because lots are made and they are all the same size.

Remember

It is important to say **why**.

A good feature of the product analysis shown in the case study is the use of '… so that …' or '… because …'. This leads into the explanation of why the product is designed the way that it is.

Activities

1 Carry out a product analysis of a chair or stool in your classroom.

2 Examine the product analysis in the case study. What extra information or explanations could have been included to make it better?

Summary

● Product analysis is about understanding why existing products are designed the way that they are.

● A product analysis can help you to make your own designs better.

● ACCESS FM is a way of remembering the things to investigate during a product analysis.

1.5 Specification

Objectives

- Understand what a specification is.
- Be able to create a specification.

Key terms

Specification: a list of needs that the product must meet.

Link

See **1.2 Understanding the design brief** for more information on analysing the design brief.

What is a specification?

The **specification** is a list of all the needs that the product must meet. Most of these are the answers to the questions you identified when you analysed the design brief.

The specification document is a very important part of the design process. It states what you can and can't do when you are designing the product. If any important needs are missed out of the specification, the product you design might not do what users need it to do, or you might not be able to make it.

Constraints

It is especially important that any constraints are listed in the specification. For example, if you only have wood available, this would be a constraint. In which case, it would be wrong to design a product that had to be made from plastic or metal.

Not all the needs in the specification will be constraints. Some will be features that users might want or things that are needed for the product to carry out the things it has to do.

ACCESS FM

A good specification for a simple product might list between 10 and 20 needs. For a complicated product, like a car, it might contain thousands of needs. ACCESS FM can be used to make sure that you have included all the different types of needs in your specification.

Aesthetics	What shape is it? What colour is it? What is the surface texture like?
Customer	Who is it being designed for?
Cost	How much should it cost to make?
Environment	Should it be made from recycled materials? Must it be able to be recycled?
Size	What size is it? (Hint: write down the sizes. For example, 'It must be less than 50 mm long' or 'It must be between 40 and 60 mm wide'.)
Safety	What things must be included in the design to protect the user from harm when using it? Does it have to meet any safety standards and regulations?
Function	How will it be used? What maintenance activities can be carried out?
Materials and manufacture	What materials can it be made from? What processes have to be used to make it? How are the parts joined together?

The specification will be used many times during later stages in the design process. For example:

- If there is a choice of possible designs, they might be compared against the specification to see which one is the best.
- When the final product has been completed, the design will be compared to the specification to check that it does what it is needed to do.

It makes it a lot easier at those stages if the needs in the specification can be tested. The result of the testing might be a yes or no answer, or it might be a measurement.

Link

See **1.4 Product analysis** for more information on ACCESS FM.

Link

See **6.1 Evaluation of the finished product** for more on comparing the product to the specification.

Case study

Educational toy for a child

This is an example of a specification for a child's toy:

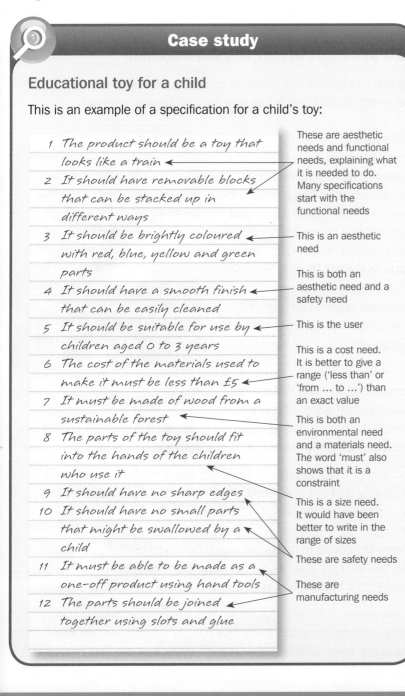

1 The product should be a toy that looks like a train
2 It should have removable blocks that can be stacked up in different ways
3 It should be brightly coloured with red, blue, yellow and green parts
4 It should have a smooth finish that can be easily cleaned
5 It should be suitable for use by children aged 0 to 3 years
6 The cost of the materials used to make it must be less than £5
7 It must be made of wood from a sustainable forest
8 The parts of the toy should fit into the hands of the children who use it
9 It should have no sharp edges
10 It should have no small parts that might be swallowed by a child
11 It must be able to be made as a one-off product using hand tools
12 The parts should be joined together using slots and glue

These are aesthetic needs and functional needs, explaining what it is needed to do. Many specifications start with the functional needs

This is an aesthetic need

This is both an aesthetic need and a safety need

This is the user

This is a cost need. It is better to give a range ('less than' or 'from … to …') than an exact value

This is both an environmental need and a materials need. The word 'must' also shows that it is a constraint

This is a size need. It would have been better to write in the range of sizes

These are safety needs

These are manufacturing needs

Activities

1 Using ACCESS FM, create the specification that you think might have been used for a desk in your classroom. For each need, include a sentence explaining why it is important.

2 Examine the specification in the case study. For each of the listed needs, write down why it might be important. What extra details could have been included to make it better?

Summary

- A specification is a list of needs that a product must meet. It should include any design constraints.

- ACCESS FM can be used to check that you have covered all the different types of need in your specification.

- It is useful if the needs in your specification can be tested.

2.1

Environmental concerns

Objectives

● **Understand why products have an impact on the environment.**

Key terms

Non-renewable: something that is not replaced and will eventually run out.

Pollution: contamination of the environment.

Sustainable materials: materials that are easily available and can be harvested, manufactured and replaced using very little energy.

The environment is the world we live in. We are used to a certain 'quality of life'. We have good food and shelter, and luxuries such as cars, phones, televisions, computers and air travel.

To make sure that we still have the same quality of life in the future, we need to protect our environment from being damaged too much by what we do.

How products affect the environment

We use lots of different products every day. However, we often forget that every product made has an effect on the environment.

Materials

Products are made from materials. Most materials are **non-renewable**. This means that there is only a certain amount of that material on our planet, and it could run out. If a certain material starts to run out, then it becomes more difficult to find and more expensive to buy. This makes it more difficult to use that material in products.

A Oil is a non-renewable resource used to make many types of plastic

The materials you can see in a product are normally not the only materials that were used when making it. There is often waste. For example, if a shape is cut from a sheet of metal, the metal that is cut away will be waste.

Remember that materials are used to make machines, tools and packaging that are then used in the making of the product as well.

Energy

Energy is needed to power the machines used to make the products and the vehicles used to deliver them. Currently, most of this energy comes from burning non-renewable materials such as oil or coal.

B A wind power generator

In the future, it is likely that more of our energy will be generated by using wind power, solar power (the sun) or tidal power (the sea). These are renewable sources, which means that they will not run out.

Pollution

Pollution is when we contaminate the environment.

When oil and coal are used to make energy, they make chemicals and gases at the same time. These are known as **by-products**. They include carbon dioxide, which is a cause of global warming. Making products from some materials can also produce unwanted by-products such as chemicals. If by-products escape into the environment, they can cause pollution.

Pollution can also be caused by throwing away products after we have finished with them. In the UK alone, we fill enough bins each year to go all the way to the moon and back. Some of this waste is burnt and a lot is buried in the ground in 'landfill' sites.

C Pollution

Case study

Batteries

Batteries are often used in MP3 players and other electrical goods. However, many batteries contain chemicals that cause pollution when they are made and when they are thrown away.

An MP3 player that uses a rechargeable battery reduces the pollution caused by making lots of batteries and throwing them away. Therefore it has less impact on the environment.

How can designers make things better?

When designing a product, it is important to consider all the different ways that the product will affect the environment – during its manufacture, while it is being used and after its disposal.

Where possible, a designer might think about using sustainable materials. These are materials that can be naturally replaced. For example, many wood products now carry the Forest Stewardship Council logo (Figure D) to show that the wood comes from renewable, managed forests. Using logos like this allows the customer to make choices about what product to buy.

D The Forest Stewardship Council logo

Activities

1 Action figures and dolls are often made from plastic and packaged in cardboard and plastic. Produce a word web that analyses all the different ways that this product might affect the environment.

2 Choose a product that you are familiar with, like a chair or a mobile phone. Write an article for a magazine that explains all the ways that the product might affect the environment.

Summary

- Every product made has an effect on the environment.

- Making products uses materials, both for the product and for the waste made when making it.

- Making products uses energy. This is often obtained from non-renewable resources.

- By-products from the making process and the disposal of products may cause pollution.

2.2

The six Rs

Objectives

- Be able to list the six Rs.
- Be able to explain what is meant by reduce, rethink and refuse.

Link

See **2.1 Environmental concerns** for more information on sustainable materials.

Products are made from parts and components. In an ideal world, these parts would all be made from sustainable materials like wood, paper or cotton. Unfortunately, for many products those materials are not suitable. One compromise is to use as little new, non-sustainable material as possible.

Thinking about the six Rs will help the designer to make sure that they use as little new, non-sustainable material as possible to make the product.

Case study

How the six Rs might be considered when designing a car

Reduce: using fewer materials can make the car cheaper; it can also make it weigh less, so it uses less fuel.

Rethink: designers might decide that different types of motor can be used, such as electric power rather than petrol.

Refuse: users might refuse to buy cars that use a lot of fuel.

Recycle: the materials that are used might be recycled – at the end of its life, the car might be melted down and used to make new cars.

Reuse: after an accident, you might be able to buy parts from a scrap yard rather than having new ones made.

Repair: most faults can be repaired. This is much cheaper than buying a new car.

The first three of the six Rs are focused on getting the most effective design for the product. These are the best approaches because they mean that you need less material for the product.

Reduce

Reduce means you should design the product so that it uses as little raw material as possible. For example, consider the supports in a bridge. This might mean calculating how strong the bridge has to be, and then calculating exactly how much material is needed so that you don't use any extra material.

Reduce could also mean using a smaller amount of a better material instead. For example, for the bridge support, you might be able to use a smaller amount of a stronger material.

Rethink

Rethink means you should think again about all aspects of the product to see if they could be made more environmentally friendly. For example:

- Is the product really needed?
- Could it be designed in a different way so that it uses less material?
- Does it need to look the way it does?
- Does it need to be packaged the way it is?

Refuse

Refuse means you should not accept aspects of a design that are not the best option for the environment. For example, this might mean refusing to use extra packaging just because it looks nicer. It might also mean that when you need to dispose of a product, you will not choose the easiest way of doing so if it is bad for the environment.

Key terms

Reduce: use fewer raw materials.

Rethink: find other ways of designing the product to make it more environmentally friendly.

Refuse: do not accept designs that are not the best option for the environment.

Link

See **2.3 The six Rs continued** for more information on the six Rs.

Activities

1 Think about a chair that will be used in a classroom:

   a Create a sketch of a similar chair that uses less material (**reduce**).

   b **Rethink** the chair – is there another way of allowing people to work? Create some labelled sketches with your ideas.

2 Make a list of any packaging that you have used today – this might include food wrappers, boxes and bags. For each item on the list, write down an alternative that could have been used if you had **refused** to use it.

Summary

- The six Rs are reduce, rethink, refuse, recycle, reuse and repair.
- Reduce means using less material.
- Rethink means reconsidering all aspects of the design.
- Refuse means not accepting things that are not the best solution for the environment.

The six Rs continued

Activity

1 Try to collect examples of plastic products with different numbers in their recycling triangle.

Recycle, reuse and repair are about making sure that once you have a design for a product, you make the most of the materials that have already been used and processed.

Recycling

Recycling means that you take the material used for a part, break it down or melt it and use it to make a new part. Recycling can be carried out for metals, glass and many different types of plastic. Products containing materials that can be recycled are often marked to show the material that they are made from (Figure **A**).

It is easiest to recycle products that are made from only one material. Unfortunately, many products are made up of lots of different parts and materials, which means that they need to be taken apart to allow recycling. As a result, it can sometimes cost more to recycle than it would to make new material.

Reuse

Reuse means using the part again. It is better than recycling because you don't have to use energy to change the shape or size of the part. For example, people use scrap yards to buy parts to repair cars. This is because it is cheaper to buy parts second hand than to buy them new.

Repair

Repair means mending parts so that they will last longer. This means that fewer new materials are needed to make replacement parts. For products like cars, it is also much cheaper to repair the car than to make a new one.

Product is made from aluminium and can be recycled

Product is made from glass and can be recycled

Product contains plastic materials that can be recycled

Product contains a certain type of plastic that can be recycled, indicated by the number and letters

A Recycling symbols

Rather than be glued together, a remote control is held together by screws, so it can be opened up to replace broken parts

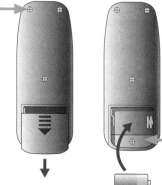

An access panel allows a new battery to be put in

B The need for repair has to be considered during the design of a product such as a remote control

Case study

Cups

Disposable plastic cups are designed to be used once and thrown away. They are designed to be lightweight, so they use as little material as possible (**reduce**). The material used can normally be **recycled**. The advantages of these cups are that they are cheap and convenient to use.

Cups made from china and ceramics can be **reused** many times, so long as they are washed! If the handle is damaged, it can sometimes be stuck back on with adhesive. This **repair** would mean that the cup could last much longer. It takes a lot of energy and material to make a cup like this, but because it can be used so many times it can be environmentally friendly.

Cups for use by campers or soldiers are sometimes made of metal. Why do you think this is?

Activities

2 Two other materials are often used to make cups:

- Reusable paper cups are coated with a thin layer of plastic to stop them being damaged by the liquid inside the cup.
- Aluminium is a type of metal that is sometimes used to make cups for use by people who are camping or hiking across the countryside.

For each of the six Rs, explain how each R has been considered during the design of cups using these materials.

3 Examine a mobile phone. Are there any symbols to show that parts of it can be recycled? Are there any parts that can be reused? What evidence is there of design to allow for repair?

Key terms

Recycling: breaking or melting down the material so that it can be used in a new product.

Reuse: using the product again.

Repair: mending a product so that it lasts longer.

Link

See **2.2 The six Rs** for more information on the six Rs.

Summary

- Recycling, reusing and repairing help to minimise the amount of new 'non-sustainable' materials used.

- Recycling involves breaking or melting the material down and making it into a new product. Not all materials and products can be recycled.

- Reusing components means that no energy is needed to change the shape of the part.

- Repairing increases the usable life of a product.

3.1

Social and cultural issues

Objectives

- Be able to give examples of how social and cultural issues may affect the design of a product.

When designing a product, designers don't just think about their product. They also have to make **moral choices** about a wide range of issues. They have to think about what is 'good' and what is 'bad' for society, and how this affects the design of the product.

How designers decide this depends on the values of society and on the market needs, such as cost. There is often no 'correct' answer and it comes down to a moral judgement.

Social influences on design

Does the product affect people other than the user?

The designer often has to balance what the product is intended to do for the user with its effect on other people. For example, an MP3 player with headphones allows the user to listen to their own choice of music. However, some people think that the noise they can hear when someone else is using headphones is very irritating.

Do we make the product environmentally friendly?

Sustainable materials sometimes cost more than non-sustainable materials. Similarly, it might cost more to design and make a product so that it can be easily recycled once you have finished with it. However, some people might buy a more expensive product if they know it is more environmentally friendly.

Link

See **2.1 Environmental concerns** for more information on the environment.

Do we design for everybody or just certain people?

If products are designed only for the 'average' user, many members of society may find it difficult to use them. For example:

- If the only difference between the green start and red stop buttons on a machine is the colour, people who are colour blind or visually impaired might get them confused.
- Society is now made up of more and more elderly people. They often find it difficult to open containers or to turn a traditional tap on or off.
- Wheelchair users might find it difficult to reach controls and switches that are positioned too high.

Many of these difficulties can be solved through thoughtful design. However, sometimes this has an effect on the cost of the product or how well it works.

A A traditional tap

B A wrist action tap – it is much easier for elderly people to use the wrist action tap

Do we use machines to make the product?

Everyone wants cheap, good quality products. Some products are designed to be made by computer-controlled machines, as they can make things at a lower cost than a human worker. However, this means that fewer people are employed.

Where do we make the product?

Another way of making products cheaper is to make them in countries where workers get very low wages. In these countries, the conditions for workers are sometimes far below those accepted in the UK. There might also be more pollution from transporting goods all over the world.

Cultural influences on design

Culture is the way that beliefs, history, tradition and lifestyle have influenced a group of people. This varies a lot between societies, or even different groups within a society. It has a big influence on the products that the people in that group use.

For example, in China, the colour red symbolises good fortune. However, in South Africa, red is the colour of mourning. Trying to sell the same red product in those two countries might get a very different response.

C Child labour

Case study

Dining in a Japanese home

In a Japanese home, it is traditional that people eat whilst sitting on a cushion or the floor. This means that the dining table needs to have very short legs and the 'chairs' may not have legs at all! Anyone designing furniture to be sold in Japan would have to take into account this cultural influence.

Can you think of other cultural differences between different countries that affect what products they use?

D A Japanese dining table

Summary

- Designers have to make moral choices about how social issues affect the design of their products.
- These choices will depend on the values of society and market influences, such as the acceptable cost.
- Social issues include how the product affects other people, the environment, employment and manufacturing location.
- Culture can have a big influence on the products that a society may use.

Activity

1 Examine a mobile phone. Write a list of all the moral issues that the designer may have had to consider when designing it. What decision do you think was made for each issue? Explain your answers.

3.2 Laws and standards

Objectives

- Be able to explain the role of laws and standards during the design of a product.
- Be able to explain the role of a patent.

Key terms

Standard: a document published by the BSI that lists all of the properties expected of a product and the tests that should be carried out.

Patent: a legal protection for a design idea.

A The BSI Kitemark

B The 'CE' marking symbol

Making products safe

Society needs to protect people from being harmed by dangerous or faulty products. It does this through laws and standards. The designer has to take these into account when designing a product.

Laws

There are laws to protect people from being injured by products. There are also laws to stop people being sold products that don't do what they are supposed to do. If a designer or manufacturer breaks these laws, then they can be taken to court and prosecuted.

The laws include:

- the **Sale of Goods Act**. This states that goods must be fit to do what they are intended to do. For example, a shop couldn't sell raincoats that dissolved in the rain!
- the **Trade Description Act**. This makes it illegal to say that the product will do something that it can't do. For example, you couldn't claim that a red t-shirt was black and you couldn't sell a vacuum cleaner that would not pick up dust.
- the **Consumer Protection Act**. This stops people selling products that might be faulty or dangerous. For example, a teddy bear for small children might have plastic stick-on eyes. If the eyes could come off and be swallowed, it would be illegal to sell it.
- the **Consumer Safety Act**. This allows the government to ban the sale of dangerous products.

Standards

A **standard** tells you what properties a certain type of product should have, and how they should be tested. For example, it might say: 'It should have a strength of more than …' or 'The size should be between …'.

There are thousands of standards, covering lots of different types of product. In the UK, standards are published by the British Standards Institution (BSI). If a product meets the needs of the relevant standard, it can be awarded the BSI Kitemark (Figure **A**). There are also European Standards. Products that meet these are also marked with a symbol (Figure **B**).

Protecting the designer's work

There are laws that protect designers and companies by making sure that their ideas cannot be stolen.

A **patent** is a legal protection that gives exclusive rights to the use of a design. This means that only the person who has the patent can use that design idea. Other people cannot use it without the permission of the patent holder. Sometimes they have to pay a fee to get this permission. Patents normally last for 20 years.

Patents help designers by making sure that they get the benefit from their ideas. When you are designing a product, make sure that you don't use features or ideas that someone else has patented.

Case study

The Dyson vacuum cleaner

Traditionally, vacuum cleaners collected dust and dirt in disposable bags. However, the more dirt that was collected in the bag, the less well the vacuum cleaner worked.

James Dyson got frustrated by this and decided to invent a vacuum cleaner that didn't need a bag. It took five years of work and 5,127 models of different ideas before he arrived at his final design. To make sure that he got the rewards for his effort, Dyson got patents for the ideas he put in his design.

The Dyson vacuum cleaner was a huge success and lots of them have been sold. A lot of other vacuum cleaner companies started making similar vacuum cleaners that didn't need bags. Dyson took these companies to court as they were using some of his ideas that he had patented. The other companies had to stop making the bagless vacuum cleaners. This meant that Dyson's company could get the maximum amount of sales from his idea.

C The inventor James Dyson and his bagless cyclone vacuum cleaner

Can you think of any other products with unique features that a designer might have tried to patent?

Activities

1 Your family has bought a chair for use in the garden. The first time that someone sat on the chair, two of the legs bent and the chair collapsed. Write a letter to the manufacturer explaining why they should give you your money back, referring to the relevant laws.

2 Patents are often listed on products, along with their reference number. Examine several products and identify those which are protected by patents or have BSI Kitemarks and CE marks.

Summary

● Designers have to think about laws, standards and other designers' patents when they are creating their designs.

● Standards list the properties expected of a product and how they should be tested.

● Patents give legal protection to a design idea to stop people using the idea without the designer's permission.

4.1

Types of drawing

A lot of different types of drawing are used in technology. Drawings are a way of communicating information. Different types of drawing are used to communicate different things.

Objectives

- Be aware of the different types of drawing used by designers.
- Know what they are used for.

Sketching

Sketching is a very effective way of capturing ideas and sharing them with other people. Sketches are normally drawn by hand, as this is quicker than using a computer. Labels are used to point out important features (Figure A). Colour or shading might be added to improve the presentation.

Key terms

Orthographic drawing: a working drawing that shows the dimensions of the part.

Exploded view: a drawing that shows how the parts of a product fit together.

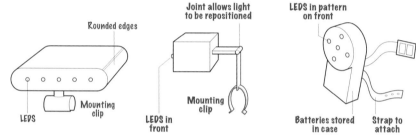

A Examples of simple sketches: ideas for a bicycle light

Isometric drawing

A designer might use isometric drawings to show ideas to his client. They provide a better three-dimensional (3-D) representation of the object being drawn than many other drawing techniques. They can be produced by hand or using computer software. They often include colour to make them look even more realistic (Figure B).

Orthographic drawing

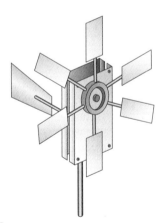

B Example of an isometric drawing: idea for a wind power generator

Orthographic drawings are also known as working drawings. Their job is to communicate the sizes of a part, so they are very useful to the people who are making the parts. They show three different views of the same product. Most orthographic drawings are produced using computer software (Figure C).

 Link

See **4.2 Sketching and crating** for more information on sketching.

See **4.3 Isometric drawing** for more information on isometric drawing.

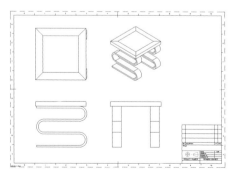

C Orthographic drawing of a table

Exploded views

Exploded views are very useful to people who are assembling a product made from lots of different parts. They are often included in the instructions for furniture that users have to put together themselves. They show all the different parts lined up, so that users can see how the product fits together (Figure **D**).

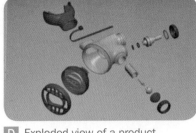

D Exploded view of a product

3-D models

One of the advantages of using computer software to prepare drawings is that you can draw parts separately and then put them together virtually to check that they all fit. You can also use other software to test the design, for example to see how strong a part would be when in use (Figure **E**).

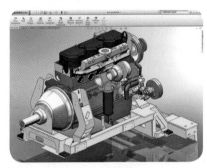

E 3-D model of a motor

Case study

Vacuum cleaner design at Dyson

The designer starts by sketching out ideas (Figure **F**). They choose the best idea and then the individual parts are each drawn as a 3-D model (Figure **G**). Computer software is then used to test how well these parts work (Figure **H**). Once they have a design that works well, Dyson builds and tests a working prototype (Figure **I**).

F Sketching initial ideas

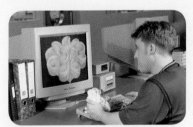

G Producing a 3-D model

H Virtual testing of a component in the Dyson vacuum cleaner

I Testing prototypes

J James Dyson's DC01 vacuum cleaner

 Link

See **4.6 Modelling techniques** for more information on 3-D modelling.

Activity

By hand, create an exploded view of a ballpoint pen. Remember to line the parts up so that you can see how they fit together.

Summary

● Drawings are a way of communicating information.

● Different drawings are used to communicate different things. For example, sketches are used to show ideas and orthographic drawings are used to communicate sizes.

4.2 Sketching and crating

Objectives

- Be able to explain how sketching is used.
- Be able to use crating to make a 3-D sketch.

What is sketching?

Sketching is a freehand drawing technique. Sketches do not have to be produced to **scale**. However, they do need to be in proportion.

Sketches can be two dimensional (2-D) or three dimensional (3-D). 3-D sketches are often used to show the whole design, with 2-D sketches used to show close-up views of details on the design.

The main aims of sketching are to see what your ideas look like and to share those ideas with other people. Sketches are often used to get your first ideas down on paper (Figure **A**). This is sometimes known as capturing ideas or producing concept drawings.

The only equipment needed for freehand sketching is a pencil or pen. The sketch should be produced quite quickly and lightly. However, this doesn't mean that it is rushed or unclear. If you do go wrong, you can either draw over the error or rub it out later.

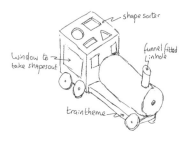
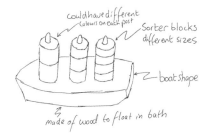
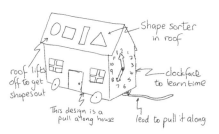

A Examples of 3-D sketches: ideas for a child's toy

Labels communicate ideas

Sketches should have lots of labels. These are very important – they explain your design thinking and help other people to understand your ideas. You might include comments about:

- where you have used ideas that you have seen in other products
- why you made it look the way it does
- constraints
- costs
- how the idea compares to your specification
- any design features due to trends or fashion
- how it could be made
- what it could be made from.

Key terms

Sketching: a quickly produced visual image of an idea.

Scale: the ratio of the size of the design in the drawing to the size of the finished item.

Crating: using a box to provide guidelines for drawing.

Crating

One way of drawing sketches in 3-D is to use **crating** (Figure **B**). This helps to get the features of your design in the correct positions and to the right proportions.

You imagine that each separate shape of the design is inside its own 3-D box, called a 'crate'. A simple shape might need just one crate. A complicated design might need several different crates for its different parts.

Follow the steps below to produce your own crated drawing:

1 First, draw the crates very lightly (Figure **C**). If you cannot draw straight lines, you could use a ruler to do this.

2 Draw one side of the object on one face of the crate, using the lines of the crate as guidelines.

3 Reflect this in the opposite side of the crate and draw lines to join up the two faces.

4 After drawing, the lines for the crate can be rubbed out.

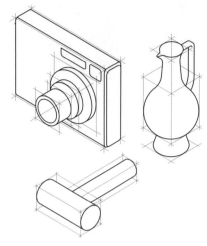

B Examples of crated drawings

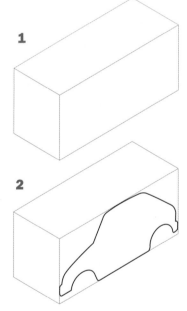

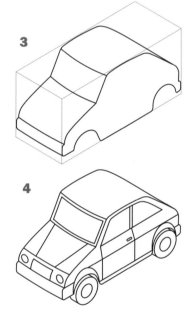

C Crating

Summary

● Sketches are used to see what your ideas look like and to share them with people. The important features of the sketch should be labelled.

● Crating can be used to provide guidelines for 3-D sketches and drawings.

Activities

1 A mobile phone company is thinking of selling a new phone to teenagers. Produce a series of quick sketches of different designs of mobile phones that the company might consider. This should include designs with touch screens, traditional keypads and a flip-top phone. Label all the important features of your ideas.

2 Using crating, produce a 3-D sketch of a lorry. (Hint: each wheel should use a separate crate.)

4.3

Isometric drawing

Objectives

- Be able to produce an isometric drawing.

Key terms

Isometric drawing: a 3-D drawing technique where horizontal lines are at 30° to the horizon.

Link

See **4.2 Sketching and crating** for more information on using labelling to communicate ideas.

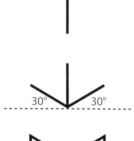

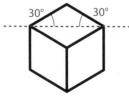

B Creating an isometric drawing

What is isometric drawing?

Isometric drawing is the most popular technique used to create 3-D drawings. Compared to drawings produced using other techniques, isometric drawings normally look more like the object. This is because the different features of the object are all in proportion to each other. Isometric drawings are often used during the design process to share ideas with other people.

The main features of an isometric drawing are:

- the closest part of the object to the person looking at it is a corner
- all horizontal lines are at an angle of 30° to the horizontal (Figure **A**)
- all the vertical lines remain vertical
- all measurements are to scale.

As for other drawing techniques, isometric drawings should also be labelled to explain important features.

Isometric paper is available to provide guidelines for isometric drawing, as used in Figure **A**.

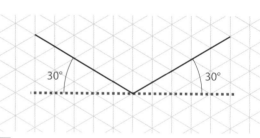

A Angle used for isometric drawing

How to produce an isometric drawing

To produce an isometric drawing, you normally start by drawing the front corner. This is vertical, so it will point straight up. You then extend this out to become the full image, remembering to keep all of the lines to scale. For example, if you wanted to draw a simple cube, you would use the following steps (Figure **B**):

1. Look at the object you want to draw from a corner.
2. Draw the corner vertical line. Remember that this should be in scale to the object being drawn. For example, if the cube is 100 mm high and the scale is 10 : 1, this line would be 10 mm long.
3. Add any horizontal lines that extend from the corner line, at 30° to the horizontal.
4. Add any vertical lines extending from these horizontal lines.
5. Add any horizontal lines extending from these vertical lines and so on, repeating steps 4 and 5 as needed to build up the complete drawing.

Drawing curves using isometric drawing

The approach described on the previous page will allow you to create excellent isometric drawings if all the sides of the object you are drawing are straight. However, design ideas often include curves or circles. Fortunately, if you can draw an isometric cube, you can use this to draw almost every other shape.

In an isometric drawing, any curve or circle will look distorted because you are looking at it from an angle (Figure **C**). This means that curves and angles cannot be drawn with compasses. There are special isometric templates that allow you to draw circles. However, if you do not have one, or the circle is of a different size to the template, another way of drawing them is to use a crate (Figure **D**).

First, you draw an isometric box called a crate. Ideally, the lines for the crate should be very faint. It is then marked where the circle will touch the edges and these points are joined up. After drawing, the lines for the crate should be rubbed out. Shapes with curved features are drawn in the same way.

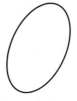
Isometric view

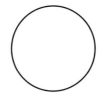
'Direct' view

C Angle used for isometric drawing

Link

See **4.2 Sketching and crating** for more information on crates.

Activities

1 Produce isometric drawings of a television and a remote control. Remember to label all the important features.

2 Produce an isometric drawing of a sports car. Remember to use an isometric crate for the wheels and any curves.

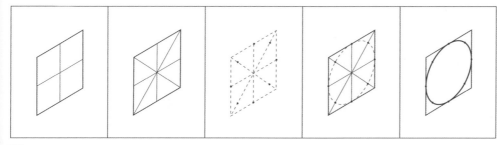

D Producing an isometric view of a circle

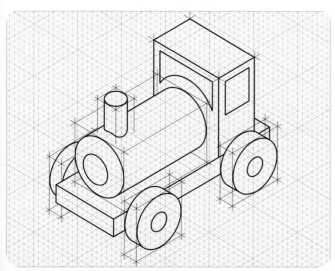

E Example of an isometric drawing – a toy train

Summary

● Isometric drawing is a 3-D drawing technique.

● On an isometric drawing, the horizontal features of the product are shown at 30° to the horizontal. Vertical features remain vertical.

● Isometric crates can be used to give guidelines for circular or curved shapes.

4.4 Rendering

Objectives

- Be able to render a sketch or drawing using different line thicknesses and shading.

What is rendering?

Rendering means adding colour or texture to a picture. The aim of rendering is to make the drawing or sketch look more realistic, so you can get a good idea of what it would look like if it was made into a finished product. Two easy-to-use forms of rendering are **thick and thin lines** and **shading**.

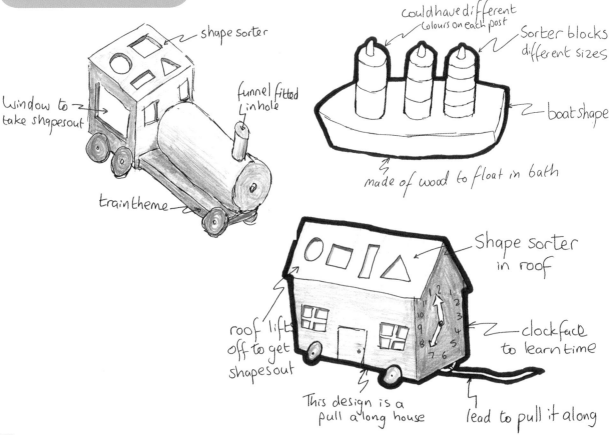

A Examples of rendered drawings: ideas for a child's toy

Key terms

Rendering: applying colour or texture to a sketch or drawing.

Tone: lighter and darker versions of a colour.

Shading: creating different tones on a sketch or drawing.

Thick and thin lines

One way to make different parts of a picture stand out is to use different line thicknesses. A simple technique is as follows:

1 Produce the drawing as normal. All the lines are thin and the same thickness (Figure **B** part 1).
2 On any edge where only one surface can be seen, the line thickness is then increased to medium (about double the thickness of 'thin' lines) (Figure **B** part 2).
3 On any outside edge, the line is made even thicker (Figure **B** part 3).

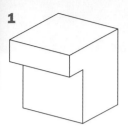
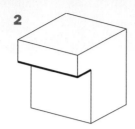
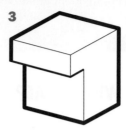

B Thick and thin lines

It is best to use technical pens, which are available in different thicknesses, such as 0.25 mm, 0.5 mm and 1 mm. The drawing can also be done in pencil, but you need to take care to keep each line thickness consistent.

Shading

If an object is placed near a window or light, the side facing the window will appear to be a lighter colour than the side that is in shade. These lighter and darker versions of the same colour are called **tones**. A drawing shaded with different tones looks more like a real product. With round objects, such as balls or pipes, **shading** also helps to make the surface appear curved.

Start by identifying the light source – where the light falling on to the object is coming from. Using a pencil, carry out the following:

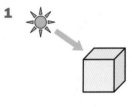
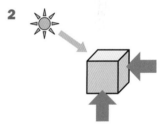
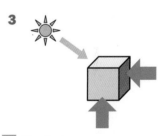

1 Lightly shade the whole object (Figure **C** part 1). This should be the tone of the lightest area, which the light is shining directly on to.
2 Identify any sides or areas that are further from the light source or that the light does not shine directly on to. Shade these again, so that they become a bit darker (Figure **C** part 2).
3 Identify any sides that have no direct exposure to the light source. Shade these again, so that they become even darker (Figure **C** part 3).

Shading is most effective if just one colour is used, with lighter and darker tones. Shading can be made even more effective by showing a shadow from the object on the opposite side from the light source.

C Shading

Activities

1 Produce three quick sketches of the same house. On one sketch use 'thick and thin lines'. On a second sketch use 'shading'. Leave the third sketch as it was first drawn. Explain which of the sketches looks most realistic.

2 Produce an isometric drawing of a bus. Use both 2-D and 3-D. Render your drawing to make it look more realistic.

Summary

● Sketches and drawings can be made to look more realistic by rendering.

● Rendering techniques include thick and thin lines and shading.

4.5 CAD

Objectives

- Be able to explain what is meant by CAD.
- Know the main advantages and disadvantages of using CAD drawing software.

Key terms

Computer-aided design (CAD): the use of computer software to support the design of a product.

Features: details on the design.

 Link

See **4.6 Modelling techniques** for more information on using CAD to test a product.

What is CAD?

Computer-aided design (CAD) is where a designer uses computer software to help design a product. CAD is most commonly used for drawing, although it does have other uses as well.

Using CAD for drawing

You will probably have some CAD drawing software in school, such as Techsoft 2-D design, SketchUp, ProDESKTOP, ProENGINEER, AutoCAD or Solidworks.

The basic commands used to create CAD drawings are called drawing tools. Most are activated by using a mouse. These tools range from drawing simple lines and inserting simple shapes, to copying, manipulating and altering the **features** on the drawing.

When you start drawing using CAD software it is a bit like drawing by hand. The screen has a working area, a bit like a piece of paper that you would draw on. This area can be changed to almost any size, and you can zoom in to see features close up. However, with 3-D CAD software, you are not limited to just drawing on the paper, you can also draw into the space above and below it! This means that you can create full 3-D models of the item that you are drawing.

The advantages of using CAD for drawing

CAD drawing has many advantages over drawing by hand:

- It can be quicker to create a new drawing, as you can copy and edit a similar drawing.
- It is easier and quicker to make changes to a drawing. To make changes to a drawing made by hand, you normally have to restart the drawing from scratch. To make changes to a CAD drawing, you can open and edit the existing file.
- Commonly used parts, such as screws, nuts and bolts, can be downloaded from libraries of CAD parts. This means that they don't have to be drawn, which saves lots of time.
- CAD drawings can be more accurate.
- CAD drawings can be saved electronically, saving space.
- CAD drawings can be easily circulated to anyone who needs them, using CDs, email or downloads from the internet.
- CAD drawings can sometimes be used to test a virtual model of a product, saving the cost of making an actual product just to test it works.

The disadvantages of using CAD for drawing

Although there are many advantages of using CAD for drawing, there are some disadvantages too:

- In industry, CAD software can be expensive and specialist training is often needed to be able to use it.
- It is harder to keep the drawings safe from competitors, as electronic files can be easily copied.
- Work can be interrupted or destroyed by computer viruses, corrupt files and power cuts. For this reason, regular back-ups should be taken of all important files.

A Creating a CAD drawing

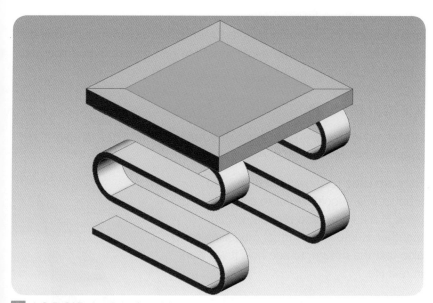

B A 3-D CAD drawing of a table

Activities

1 Using the CAD software available in your school, produce a drawing of your classroom.

2 A local company makes furniture. The range of products that the company makes changes every year, depending on fashion. The manager of the company has decided that he does not want the cost of buying new CAD software, so he is thinking about getting workers to produce any drawings they need by hand. Write the manager a letter explaining why this might be a bad decision.

Summary

- CAD is the use of software to assist in the design of a product. This includes the use of CAD software to produce drawings.

- Compared to drawing by hand, CAD drawings are easier to change, can be more accurate and are easier to store and distribute.

Modelling techniques

Objectives

- Understand the reasons for modelling a design.
- Be able to list different ways of making a model.

Key terms

Model: a representation of a design idea.

Soft modelling: making a model using materials that are different to the final product.

A Modelling a lampshade using card and paper

Why make a model?

Making models of your design is an important part of the design process. A **model** is a representation of your design. It might be of just one feature or it might be of the whole design. Models have many uses:

- You can test out how well different design ideas work.
- Models give you an impression of what the finished item will look like.
- You can test whether features, such as moving parts, work the way you want them to.
- You can identify where it might be difficult to make the design.
- You can quickly make changes to your design to improve it.

Modelling helps to make sure that the finished product is correct the first time that it is made. It is normally much cheaper to make a model than to make a finished product.

You could make a finished product without modelling. However, if it didn't work or you identified improvements, you would have to make it again. This might cost you a lot more time and money than if you had made a model first.

Virtual models

Virtual models only exist inside computer software. For example, they could be 2-D or 3-D CAD drawings.

They are used to show what a finished product might look like, or to see if the parts in a complicated product will fit together. Sometimes, virtual modelling can also test the design, to see how it will work in real life.

One of the main advantages of virtual models is that they save the cost of making. They can be changed quickly to check out different ideas.

Soft modelling techniques

Soft modelling means making a model using materials that are different to the final product. They might make it easier to understand what the final product will look like or how it might be made than a virtual model.

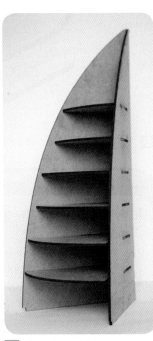

B A card model

Soft models can be made from a wide range of different materials. As these are different from the materials in the final product, the model will not have all its properties. However, it is normally quicker and cheaper to make. Soft model materials include:

- paper or card, which is often used for simple structures
- styrofoam – a type of plastic that is easy to cut and shape with a hot wire
- balsa or jelutong – hardwoods that are easy to cut, sand and shape.

If a design has been created using CAD software, it is also possible to create a model using computer-controlled machines. This is called a rapid prototype.

Link

See **8.8 CAD CAM and rapid prototyping** for more information on rapid prototypes.

Case study

Modelling vacuum cleaners at Dyson

3-D CAD models are used to draw the product and check that the parts fit together (Figure C). These can then be tested virtually using other software (Figure D). The designs are also modelled using card (Figures E and F) to see what the product might look like when it is put together.

Can you think of any other types of model that Dyson might use?

C Creating a 3-D model of a part to check the fit of components

D Virtual testing of a product – different colours represent different properties

E Making a physical model from card

F Vacuum cleaner model made partly from card.

Activity

1 When a mobile phone is not in use, one option is to put it on a stand. This is sometimes known as a 'cradle'. Sketch some ideas for a mobile phone cradle that could be made from sheets of material. Make a soft model of your favourite design using card.

2 Using the CAD software in your school, create a virtual model of a rack to hold 10 DVDs.

Summary

- A model is a representation of a design.
- Virtual models exist only within computer software. They are used to check that parts fit together or to test a product virtually.
- A wide range of different materials can be used to model products. They are used to understand what the final product will look like or how it might be made.

5.1

The production plan

A **production plan** is a set of instructions for making a product. It should contain enough information so that someone who has never seen the product is able to make it.

Why do we need a production plan?

There are lots of reasons why you should plan out what you are going to do.

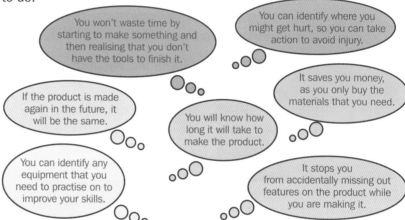

You won't waste time by starting to make something and then realising that you don't have the tools to finish it.

You can identify where you might get hurt, so you can take action to avoid injury.

If the product is made again in the future, it will be the same.

You will know how long it will take to make the product.

It saves you money, as you only buy the materials that you need.

You can identify any equipment that you need to practise on to improve your skills.

It stops you from accidentally missing out features on the product while you are making it.

Deciding how to make the product

The first thing you need to do is decide how you will make the product. This means that you need to know what material you can make the product from and what processes you will use to make it.

For example, say you want a small plastic bracket like that in Figure **A**. It has two holes and a bend in the middle. You have a large sheet of plastic of the correct thickness. You will need to:

- mark out the size of the part needed
- cut the sheet to size
- bend the sheet
- make the holes.

Next, the tasks need to be put in the right order. Some tasks will need to be carried out before others. For example, you might need to make the holes before the sheet is bent, so that it fits on the drill. The final order can be shown as a **flow chart** (Figure **B**).

Writing the production plan

On its own, the flow chart does not give enough information to make the product. Each of the tasks in the flow chart is a separate step in the production plan. For each step, there should be enough detail

Objectives

- Know why it is important to have a production plan.
- Be able to list the things that should be included in a production plan.

Key terms

Production plan: instructions on how to manufacture a product.

Flow chart: a sequence of activities presented as a diagram.

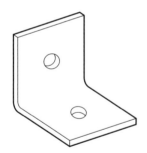

A A simple plastic bracket

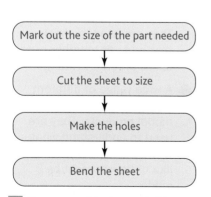

B Sequence of tasks needed to make a simple part

so that the person carrying it out is able to do it without needing any more information.

If you have a product with lots of different parts, in industry you would normally write separate plans for each part. You might also have a production plan for putting all of the parts together to make the finished product.

What should be included in the production plan

- The tasks you need to do.
- What materials to use.
- What tools to use.
- Safety notes.
- Tools to use if the first choice is not available.
- Time needed to do each task.

Step	Task	Time (minutes)	Tools to use	Materials to use	Safety notes
1	Mark out: • the size of the sheet needed (100 mm x 50 mm) • the line to show where the plate should be bent (halfway along the 100 mm length) • where the holes should be (25 mm from each end, in the middle)	15	Scribe, engineering rule	Plastic sheet, 150 mm x 100 mm x 3 mm	
2	Cut the sheet to the size needed	15	Scroll saw or a coping saw	Sheet from task 1	Use the machine guard on the scroll saw. Wear goggles
3	Make the holes, 6 mm diameter	5	Pillar drill or hand drill	Sheet from task 2	Use the machine guard and wear goggles. Use a machine vice to hold the sheet securely
4	Bend the sheet	5	Line bender	Sheet from task 3	Keep fingers away from the hot equipment and the hot plastic or they might get burned

Activity

1 Making a cup of cocoa involves the following activities (not in this order):

- Pour hot milk into cup.
- Get a cup from the shelf.
- Put two spoonfuls of cocoa in cup.
- Stir.
- Heat milk in a pan.

a Create a flow chart showing the tasks in the correct order.

b Create a production plan that can be used by someone who has never seen a cup of cocoa being made.

Summary

- The production plan gives you instructions on how to make the product.

- The production plan should include all the information needed to make the product, including the tasks to be carried out, the materials and tools to be used and safety notes.

A Example of a one-off product: a piece of designer furniture

B Example of a batch-produced product: chairs

Key terms

One-off production: making one of a product.

Batch production: making a quantity of a product before switching over to the next product.

Mass production: making the same product on an assembly line.

Continuous production: making the same product 24/7.

Products are made in different quantities. The amount of a product that needs to be made is called the **scale of production**. This normally has a big effect on the cost of the product and the type of equipment that you can use to make it. There are four commonly-used categories for scale of production.

One-off production

One-off production is where you make a single product. This is often made to an individual design for one customer. Examples of one-off products range from bronze statues to musical instruments, 'custom-made' furniture and the anchor for an aircraft carrier.

One-off parts are often made by hand by skilled workers. They might be made using hand tools or standard workshop equipment. The products are normally quite expensive because of the amount of time that the workers have to spend making them.

Batch production

Batch production makes small quantities, from a few to a few thousand, depending on the type of product. Each batch of products will have the same design. However, different batches might be customised in some way, for example the same design made in a different colour or size.

If a company is using batch production to make 100 tables, it will make 100 sets of legs, 100 table tops and then put all the parts together.

Batch production normally uses machine tools. It costs less to make products this way than with one-off manufacture because you don't have to spend as much time setting up the machines to make each product.

Mass production

Mass production makes large quantities of the same product, often thousands each day. Most things that you use every day will probably be mass produced, including cars, mobile phones and chocolate bars.

Mass production is usually carried out on an **assembly line**. This is a collection of machines that are only used to make that product. Each machine will just do one thing to the product, before passing it on to the next machine. Many of the machines will be controlled by computers and some robots might be used.

The cost of setting up a production line is very high, so you have to make large quantities of products to pay for it. However, because so many products are made, the cost of each one is normally less than for batch production.

Continuous production

Continuous production is used to make products like steel, oil or chemicals. Many of these products are used as the materials to make other products. Factories that operate continuous production often run 24 hours a day, seven days a week. The process needs to be continuous because it would be very expensive to stop it and then turn it on again.

This table shows the relationship between the scale of production, the cost of equipment needed, and the cost of each final product.

C Example of mass production: cars

	Continuous production	Mass production	Batch production	One-off production
Number of products to make	Highest ←			→ Lowest
Cost of equipment	Highest ←			→ Lowest
Product cost	Lowest →			→ Highest

D Example of a continuous production: making steel strip

Activities

1 List three examples of products that are made using each of the four different scales of production.

2 Skateboards are often made by batch manufacturing. List all the different parts in a skateboard – make sure to include any fixings, such as screws or nuts and bolts. For each part, identify its scale of production.

Summary

- The 'scale of production' is the amount of a product that needs to be made.
- As the scale of production increases from one-off to batch to mass and continuous production, more use is made of computer-controlled equipment and dedicated machines.

5.3 CAM and CIM

Objectives

- Understand what is meant by CAM and CIM.
- Understand the advantages of using CAM.

Key terms

Computer-aided manufacture (CAM): using computers to operate machine tools.

Computer-integrated manufacturing (CIM): using computers to control the entire production process, with no human input.

Activities

1 Use CAD to design a nameplate for your room and then manufacture it using a CAM machine.

2 Write a short article (200 words) that could be used in a car magazine, explaining how the use of CAM to manufacture parts for cars could benefit the customer.

3 Create a cartoon strip that shows the sequence of tasks to be carried out when using CIM to design and make a simple part using a CAM machine. This should be suitable for use to explain CIM to Year 6 pupils.

What is CAM?

When using machine tools to make products, there are two ways in which they can be controlled. Manual machines are controlled by skilled workers, and CAM machines are controlled by computers.

Computer-aided manufacture (CAM) means using computers to control the machine tools. They are also called **computer numerical control (CNC)** machines, as the computer uses numbers to control them. There are probably several CAM machines in your school, such as vinyl cutters, routers and milling machines, lathes and laser cutters.

Why use CAM?

CAM has several advantages over manual machines:

- CAM machines can work faster and more accurately.
- They can do a job again and again and the product will be the same every time.
- They do not need to take breaks – they can work 24 hours a day, seven days a week.
- They can produce shapes that are difficult to manufacture using manual machines.

There are two main disadvantages to using CAM machines:

- They are more expensive than manual machines. This means that they are normally only used when there are lots of parts to be made.
- It can take a long time to write the programs to operate the machines.

CAD CAM and CIM

If a product has been drawn using CAD, computer software can analyse the drawing and create the control program (Figure **A**). This can be downloaded straight to the CAM machine to make the part. This linking of processes is referred to as CAD CAM.

Computer-integrated manufacture (CIM) is an advanced form of CAD CAM, where computers control the entire production process. CIM is a fully automated system, where individual processes exchange information with each other and initiate actions without any human input.

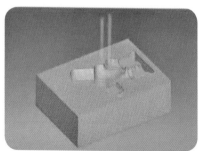

A A CAD package creating the machining procedure for a mould

Case study

Using CAD CAM at AP Valves

B Breathing apparatus for diving

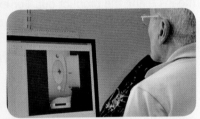

C Producing a CAD model of the component

D Making the mould using a CNC mill

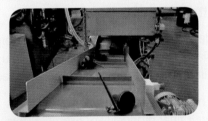

E The components leaving the injection moulding machine

AP Valves makes breathing equipment for use by divers (Figure **B**).

They draw the parts for the equipment using CAD software (Figure **C**). The parts need to be made by injecting liquid plastic into a mould. Using the drawing, the CAD software can create the image of the mould in seconds. If this had to be drawn by hand, it would take several hours. More CAD software uses the image of the mould to create the control program for a CAM machine (Figure **A**).

The mould is then manufactured using CAM (Figure **D**). After checking, the mould is used in the injection moulding machine to make the part (Figure **E**).

Summary

- CAM is the use of computers to operate machine tools.
- CAM machines can be faster, more consistent and more reliable than manual machines. However, they need time to program and are more expensive.
- CIM is a fully automated form of CAD CAM.

 ## Link

See **4.5 CAD** for more information on CAD.

6.1

Evaluation of the finished product

Objectives

- Be able to explain why it is important to compare the product to the specification.
- Be able to use evaluation to identify how the product could be improved.

Key terms

Functional testing: testing in real use, to check that the product does what it is meant to do.

Objective: based on facts, rather than opinions.

A A child's toy being functionally tested

When you have finished making your product, you need to check that it meets all of the needs you identified early in the design process.

Functional testing

The first step in evaluating the product is to check that it satisfies the design brief. One way to do this is **functional testing**. This normally means trying the product out. For example, if the design brief was to make a rack to hold 20 DVDs, you could check that 20 DVDs fit in the rack.

Comparing the product to the specification

Functional testing does not normally cover all of the needs that the product must satisfy. For example, it probably would not cover cost, aesthetic appeal or constraints for which processes could be used to make it. To evaluate these needs you should test every need in the specification. This will probably mean that a number of different tests need to be carried out.

Where possible, any testing should be **objective**. This means that it should be based on facts and numbers, rather than opinions. The table on the next page shows an evaluation against the specification for an educational toy. This has a mixture of objective tests and opinions.

Using the evaluation to improve the product

During the evaluation, you might identify some needs that have not been met. If this happens, you must explain why they have not been met. You should also identify what could be improved to allow them to be achieved.

Even if the product meets all the needs, there might be some things that you are not happy about in your final product. For example, this could be the quality of the finish or that the shape of the design made it difficult to make. This is normal – your skills and knowledge are improving as you are designing and making it.

You should note down any improvements that could be made. In industry this is very important, as the next person to make the product can use your ideas to improve the product.

Link

See **1.5 Specification** for more information on the specification.

Activity

1 Create a list of the tests that might be carried out during the evaluation of a desk for use in a classroom. Using these tests, work out what the specification for the product was.

Case study

Educational toy for a child

This is an example of testing against a specification for a child's toy:

Specification	How tested	Result of testing	Pass/fail
1 The product should be a toy that looks like a train	Visual check	It looks like a train	Pass
2 It should have removable blocks that can be stacked up in different ways	Functional testing by my three-year-old brother	He could stack the blocks up in different ways	Pass
3 It should be brightly coloured with red, blue, yellow and green parts	Visual test	It has parts of all the different colours	Pass
4 It should have a smooth finish that can be easily cleaned	Cleaning it with a cloth after my brother played with it	It could be wiped clean	Pass
5 It should be suitable for use by children aged 0 to 3 years	Functional testing by my three-year-old brother	He liked it	Pass
6 The cost of the materials used to make it must be less than £5	Added up the cost of the materials used	£4.22	Pass
7 It must be made of wood from a sustainable forest	By asking the people that supplied the wood	They said it was pine softwood from a sustainable forest	Pass
8 The parts of the toy should fit into the hands of the children who use it	Functional testing by my three-year-old brother	He could pick it up	Pass
9 It should have no sharp edges	Silk test – pulling a piece of silk over the edges and seeing if it snags	No snags	Pass
10 It should have no small parts that might be swallowed by a child	By pulling at all the small parts to make sure that they didn't come off	No parts came off	Pass
11 It must be able to be made as a one-off product using hand tools	By making it	Pass	Pass
12 The parts should be joined together using slots and glue	By making it	Pass	Pass

Activity

2 In the case study, identify which of the tests carried out were objective and which were based on opinions. For each test based on an opinion, identify either a way of testing objectively or suggest how the specification could be changed to allow objective testing of that need.

B A child's toy

Summary

● Functional testing should be carried out to make sure that the product works as intended.

● The product should be compared to the specification to check that it meets all of the identified needs.

● You should also identify how the product could be improved.

7.1   Properties of materials

Objectives

- Be able to list a range of material properties and explain how they relate to the design needs.

Key terms

Properties: how the material performs in use.

A Gears can be made from a variety of different materials depending upon the properties needed

Properties of materials

Different materials have different **properties**. This means that each material might perform differently when tested or used in the same way.

When you are designing a product, you need to choose a material that has the properties needed by your design. The first step towards doing this is to look at the design needs in your specification. These should tell you what the product you are making needs to do, the type of environment that it will be used in and how long it is expected to last. For each need, you should identify the properties that the material the product will be made from must have.

Some of the questions that you may ask and some of the material properties that you might consider are shown in the table below.

Design question	Property
Does the product have to withstand forces, such as loads being put on it, being pulled or being twisted?	Strength
Does the product need to be resistant to scratches and wear?	Hardness
Does the material need to be resistant to knocks and bumps?	Toughness
Does the material have to withstand wear and tear?	Durability
Does the product have to be light, so that it can be moved easily or carried around?	Weight
Does the product need to be in a certain price range?	Cost
Does the product have to work in an environment that could damage it?	Corrosion resistance
Does the material need to allow electricity to pass through it?	Electrical conductor
Does the material need to stop heat from passing through it?	Thermal insulator

Here are some examples:

- If you are designing a bridge over a river, you might decide that the materials used need to be strong, tough and resistant to corrosion from water.
- If you are designing a tyre for a car, you might decide that the important properties are that it should be hard (resistant to wear), tough and low cost.

- If you are designing the cover for a games console, you might decide that the important properties are that it should be an electrical insulator, lightweight and tough.

Many products need the material to have a combination of different properties. It can be hard to identify a material with exactly the right combination, so the designer often has to compromise. For example, you might choose a less attractive material because it is stronger; or you might change the design so that it uses a less strong material if it is cheaper.

Classifying materials

Materials can be classified into different types. The most common way of doing this is based on what they were made from. This gives five main categories:

- wood
- metal
- polymers, often referred to as plastics
- ceramics
- composites.

Some new materials have been developed with properties that change in response to their environment. All of these materials fall into one of the five types. However, they are often classed as a separate type called **smart** materials.

We often think that, within a type, each material will have similar properties. For example, we might believe that all metals are strong or that all polymers are flexible. While there may be some typical properties within each type, there will be big differences between individual materials. When choosing the best material for a design the material types can provide a starting point, but the designer must think about the properties of individual materials.

B All of these products have been designed to be resistant to wear

Summary

- Materials have many different properties.
- The designer needs to choose materials that meet the properties needed by the product.
- Materials are commonly classified into five types: wood, metal, polymers, ceramics and composites.
- Within each type, there may be big differences in the properties between different materials.

Activities

1 Create a list of the material properties that would be needed for the following applications:

- a chair to be used in the classroom
- a beaker to be used in a science laboratory
- a frying pan
- a chisel
- a case of a mobile phone.

2 List five examples of products made from each of the five types of material:

- wood
- metal
- polymers
- ceramics
- composites.

Wood and manufactured board

Wood is one of the oldest materials used by humans. It has been used to build houses and sheds, toys, tables, chairs and furniture.

Solid timber

There are two types of solid timber: **hardwood** and **softwood**. These names do not refer to how hard the wood is. For example, redwood is a softwood but it is hard, and balsa is a hardwood but it is soft.

Hardwoods

Hardwoods come from trees that lose their leaves in autumn and winter. These are called deciduous trees. Hardwoods include oak, beech, elm, ash, mahogany, teak and balsa wood.

A Deciduous tree

B Coniferous tree

Softwoods

Softwoods come from trees that keep their leaves all year round. These are called coniferous trees and include pine trees, which are also used as Christmas trees. In general, they grow faster than deciduous trees. Softwoods can be grown in sustainable forests, meaning that more wood can quickly be grown to replace any that is used.

Conversion

Conversion means cutting the trunk of the tree into planks. It takes a lot of time to cut a tree trunk into planks of wood. The planks are normally fairly narrow – they can never be wider than the size of the tree!

The properties of a plank can vary in different directions, depending upon its **grain**. The grain is the direction of the fibres that make up the wood. This can be seen as the different layers within the wood.

Manufactured boards

Manufactured boards are made from wood. Often they are made from the waste or off-cuts from cutting solid timber. They tend to be cheaper than solid wood planks.

Manufactured boards are available in larger widths than timber. As they have no grain, their properties can be uniform in different directions. They are often covered with a thin layer of timber, called a **veneer**, to improve their appearance. This means that they can be used where someone wants them to look like more expensive solid wood.

Types of manufactured board

C Structure of a tree

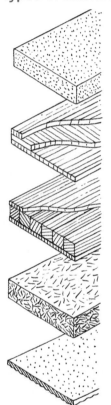

Medium density fibreboard (MDF) is made by squashing tiny particles of timber together with glue. It is a cheap flat board with equal strength in all directions. It is often used in furniture. You have probably also seen it being used to make design features on many television makeover programmes!

Plywood is made by gluing veneers at right angles to each other, to build up the thickness needed. Plywood has a constant strength and thickness. Plywood for indoor use is used to make furniture. Plywood for outdoor use is used to make signs.

Blockboard is made by gluing together strips of softwood and then covering them with a veneer on both sides. It is a strong and stable board that is often used in furniture.

Chipboard is made from thousands of chips of timber that are mixed with glue and squashed together. Compared to timber it is not as strong, but it is very cheap. It is often used with a veneer to protect it from moisture for kitchen worktops and table tops.

Hardboard is a bit like thick card, which is shiny on one side and rough on the other. It is made by squashing a 'pulp' of wood fibres mixed with glue. Hardboard is very cheap and used for cupboard backs and drawer bases.

D Types of manufactured board

Activities

1 Make a list of all the products made from wood that you used yesterday.

2 The top of a dining table might be made from mahogany (a hardwood), pine (a softwood) or veneered chipboard. Explain the advantages and disadvantages of each of these three choices.

Summary

- Wood products are available as solid timber or manufactured board.

- Hardwoods include oak, beech, mahogany and balsa wood.

- Softwoods include pine and redwood.

- Manufactured boards include plywood, blockboard and MDF.

Metals

Objectives

- Be able to explain the difference between a ferrous and a non-ferrous metal.
- Be able to give examples of different types of metal and what they are used for.

Key terms

Alloy: a mixture of two or more metals.

Ferrous metal: a metal that contains iron.

Non-ferrous metal: a metal that does not contain iron.

Corrosion resistance: the ability of a material to prevent damage due to chemical reactions with its environment.

A Metal casting

Metal has been used for many centuries because it is strong and tough. It can also be made to be shiny and attractive. Originally, it was used to make weapons, tools and jewellery. Now it is used in thousands of different types of product, ranging from bridges to cars, airplanes to fridges, cutlery to pens.

Making metals

Metal is made from metal ore, which is rock. This has to be mined from the earth. The ore is heated to high temperatures in a furnace (Figure **A**). This produces liquid metal, which is then cooled to become solid. Metals that have already been used in products can be recycled by being re-melted again in the furnace.

Most metals are not used in their pure form. Normally they are mixed with other metals to improve their properties. A mixture of two or more metals is called an **alloy**. The amount of the other metal added to form an alloy can range from 0.1 per cent up to 50 per cent. The properties of the alloy will depend upon the amounts of the different metals in it.

As they are hard and strong, it can be difficult to make metal into the shape that you want. To make this easier, metals can be bought in a wide range of shapes and sizes.

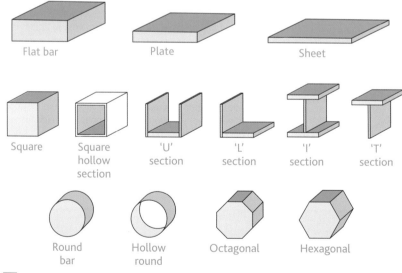

Flat bar   Plate   Sheet

Square   Square hollow section   'U' section   'L' section   'I' section   'T' section

Round bar   Hollow round   Octagonal   Hexagonal

B Shapes that metals are available in

Ferrous and non-ferrous metals

There are two types of metal alloy: **ferrous** and **non-ferrous**. Ferrous metals contain iron. Non-ferrous metals do not contain iron. Both types of metals can be recycled.

C Mild steel building frame

D Stainless steel sink

E Copper pipework

Some of the most common ferrous metals are called steel. There are many different types of steel, but all are made from iron alloyed with carbon. The amount of carbon in the steel has a big effect on the properties of the steel. Most steels are stronger than the non-ferrous metals.

In general, non-ferrous metals have better **corrosion resistance** than the ferrous metals. This means that they are less likely to rust or corrode. Some of the common non-ferrous materials include aluminium, copper and zinc. These metals are sometimes used in their pure form, but can also be used as alloys.

Activity

1 Make a list of all the products made from metal that you have used so far today. This should include metal parts used alongside other materials in products. State whether each of these products was a ferrous or non-ferrous metal, giving a reason for your answer.

Types of metal

Metal	Alloy or pure metal	Ferrous or non-ferrous	Properties	Typical uses
Mild steel	Alloy (iron with 0.15–0.35% carbon)	Ferrous	Strong, cheap, needs to be protected from rusting	Building frames, car bodies
High speed steel	Alloy (iron with 0.8–1.5% carbon)	Ferrous	Very hard and strong	Tools
Stainless steel	Alloy (iron with chrome and nickel)	Ferrous	Hard, tough, resistant to corrosion	Sinks, knives and forks, dishes
Aluminium	Pure metal	Non-ferrous	Lightweight, strong for its weight, easier to shape than steel	Cooking pans, aircraft frames, drinks cans
Copper	Pure metal	Non-ferrous	Expensive, easier to shape than iron. Electricity flows through it really well	Pipes and plumbing fittings, electrical wires
Zinc	Pure metal	Non-ferrous	Corrosion resistant. Zinc has a lower melting point than many metals, so it is easier to mould	Camera bodies, handles for car doors
Brass	Alloy (65% copper and 35% zinc)	Non-ferrous	Strong, good corrosion resistance	Castings and ornaments, door knobs

Summary

- An alloy is a mixture of two or more metals.
- Ferrous metals contain iron. Common ferrous metals include low carbon steel and stainless steel.
- Non-ferrous metals do not contain iron. Common non-ferrous metals include aluminium, copper and brass.

7.4 Polymers

Objectives

- Be able to explain the difference between a thermosetting and a thermoplastic polymer.
- Be able to give examples of different types of polymer and what they are used for.

Key terms

Polymer: the correct name for what we normally call 'plastic'.

Thermoset: a polymer that cannot change its shape when heated.

Thermoplastic: a polymer that can change its shape when heated.

B Example of a product made from thermoset: plug and socket

The correct term for what we normally call plastics is **polymers**. The first 'synthetic' polymer was Bakelite. This was invented in 1907. **Synthetic** means that it was made from chemicals. It is amazing to think that in only 100 years, hundreds of different types of polymer have been invented. Tens of thousands of different products are now made from polymers. They affect every part of our daily lives.

Types of polymers

Most polymers are made from crude oil. The oil has to be processed and a complex series of chemical reactions take place to make the polymers.

A An oil refinery

It is difficult for electricity and heat to move through most polymers, so they are good insulators. Some of the stronger polymers can be as strong as metals. Polymers are not normally painted, but their colour can be changed by adding colours to the plastic before it is made into a product.

There are two main types of polymer: thermosetting polymers and thermoplastic polymers.

Thermosetting polymers

Thermosetting polymers are also known as **thermosets**. They are normally formed into a shape by a chemical reaction between liquids or powdered ingredients. Once they have set, their shape cannot be changed again. They are harder and more rigid than thermoplastics. They cannot be recycled.

Thermoset	Properties	What it is used for
Epoxy resin	Good resistance to chemicals and wear. Strong when reinforced	Adhesives, printed circuit boards
Melamine formaldehyde	Good strength. Resistant to scratches	Laminates for work surfaces, plastic plates
Urea formaldehyde	Good strength, but brittle – will break rather than change shape	Electrical fittings, light sockets, switches, plugs

Thermoplastic polymers

The term 'plastic' that we often use for all polymers is an abbreviation of **thermoplastic** polymer.

Thermoplastic polymers soften when heated. They can be shaped when they are hot. The shape will harden when it is cooled, but can be reshaped when it is heated up again.

Most thermoplastics can be bought either as powders (that can be melted and formed into the shape needed) or as sheets of material. Thermoplastics are normally softer and more flexible than thermosets. Many thermoplastic products are marked with a symbol to show that they can be recycled. This normally includes a number to show what type of thermoplastics they are.

Thermoplastic	Properties	What it is used for
Low density polyethylene (LDPE)	Soft and flexible. Not very strong	Carrier bags, detergent bottles, packaging
High density polyethylene (HDPE)	Strong	Bowls, buckets, milk crates
Polypropylene (PP)	Good strength, but brittle – will break rather than change shape	Lunch boxes, plastic chairs, children's toys
High impact polystyrene (HIPS)	Light but strong	Vacuum formed packaging and casings
Acrylic (polymethyl methacrylate – PMMA)	Can be transparent, like glass, or coloured with pigments. Hard wearing and will not shatter on impact	Plastic windows, bath tubs, display signs
Polyvinyl chloride (PVC)	Stiff and hard wearing	Pipes, coverings for electric cables, floor and wall coverings, packaging

Link

See **2.3 The six Rs continued** for more information on recycling.

C Example of a product made from thermoplastic: a plastic bottle

Summary

- There are two types of polymers: thermosets and thermoplastics.
- Once a thermoset has been made into a product, its shape cannot be changed again.
- The shape of a thermoplastic can be changed by heating it up again.

Activity

1 You have been asked to deliver a lesson to a group of Year 6 pupils at a local primary school. The subject of the lesson is: 'What is the difference between thermosets and thermoplastics?' The lesson should be 30 minutes long and must include at least one activity for the pupils to carry out.

Write out a lesson plan that explains:

- how you would teach this lesson
- what materials or examples you would use
- any activities that you might carry out.

Objectives

- Be able to describe the properties and typical uses of ceramics.
- Be able to explain what is meant by 'composite material'.

Key terms

Ceramic: an inorganic material, normally an oxide, nitride or carbide of metal.

Inorganic: a natural material, but one that does not grow.

Composite: a material that is made from two or more material types that are not chemically joined.

A A ceramic furnace lining

Ceramics

When most people think of **ceramics**, the first things that come to mind are glasses, mugs and pots. However, ceramic materials are used for lots of different applications. These range from house bricks to plaster casts on broken limbs to the brakes in high-performance cars.

Ceramic materials are **inorganic** and non-metallic. It is difficult to make ceramic products by machine as they are very hard and can break easily. They are frequently made by moulding processes, often using high temperatures, so that machining is not required. One of the most basic examples of the 'raw material' used to make ceramic products is clay. In many parts of the world, clay is dug straight from the ground. Once shaped, it is heated to high temperatures in kilns to create the finished product. With the exception of glass, it is not normally cost effective to recycle ceramic materials.

Properties of ceramics

Ceramics have excellent resistance to corrosion and are harder than most other materials. This means that they are very difficult to scratch or wear out. However, they tend to be brittle – this means that if they are pulled they don't stretch, but tend to crack and fall apart. If they are dropped, they tend to break or shatter.

Heat and electricity find it difficult to pass through ceramics. This means that they are good insulators, so they are used as furnace linings or as heat-resistant panels on space craft. They can be made much hotter than metals without becoming soft. In fact, they are often used to make the ladles and containers used to hold liquid metal in a foundry.

Uses of ceramics

- Tools for grinding and cutting.
- Tiles to insulate furnaces.
- Electrical insulators, such as the casings on spark plugs.
- Glass equipment used in science laboratories.
- Brakes for high-performance cars.
- Glass lenses.
- Building materials, such as plaster, cement and bricks.

B Ceramic bearings

Composites

Composite materials are made by combining two or more different types of materials. This could be a combination of ceramic and plastic, ceramic and metal or even plastic and metal. Although the two materials are mixed together, they are not joined chemically. This means that they remain physically separate within the piece of composite material.

Fibreglass is a common composite material. It is often used to make car body parts or canoes. It is a mixture of ceramic glass fibres and a plastic resin. If you look at a piece of fibreglass under a microscope, you will see the individual glass fibres surrounded by the plastic (Figure C).

C Structure of a fibreglass composite – this material has been broken in two and magnified so that we can see inside. Note the glass fibres surrounded by the plastic

Properties of composites

Composites combine the properties of the materials that they are made from. For example, one of the most common composite materials is reinforced concrete. This is widely used to build office buildings and bridges. It is made by combining concrete with steel rods. Reinforced concrete has the strength of concrete with some of the toughness of steel – this means that it is much less likely to crack or break than non-reinforced concrete.

Composites can have unique combinations of properties. This means that you can make products that were not previously possible, such as lightweight bullet-proof armour. However, there is a major disadvantage: most composites cannot be recycled as it is very difficult to separate the two materials.

D Canoes made from composite materials

Uses of composites
- Reinforced concrete.
- Buildings and construction.
- Fibreglass.
- Boat hulls, canoes, vehicle body panels.
- Carbon reinforced plastic (CRP).
- Racing car bodies, helmets, armour.
- Metal matrix composites.
- Engine parts in high-performance cars.

E A bicycle with a frame made from composite material

Summary

- Ceramic materials are amongst the hardest engineering materials, but they tend to be brittle. They have excellent corrosion resistance and are good insulators to both heat and electricity.

- Composites are a combination of two or more different types of materials. They combine the properties of the materials that they are made from.

Activities

1 Choose a product made from either a ceramic material or a composite material and carry out a product analysis. This should include an explanation of why this material was used to make the product.

2 Recommend a suitable material to make the frame for a high-performance sports bicycle. Explain your choice by comparing it to an aluminium frame.

7.6 Smart materials

A smart material has a property that changes when its environment changes. This means that it has a property that can be changed by, for example, temperature, light or pressure. The change can happen again and again. For example, you've probably seen spectacles with lenses that get darker as the light gets brighter. When the spectacles are taken out of the bright light, the lenses will become less dark again.

Although smart materials are sometimes treated as a separate category of material, all smart materials fit into other material categories, such as metals or plastics, for example.

Smart materials give designers some really exciting options to consider for new products. The examples below are just a small selection of what is available and many more are being invented.

Shape memory alloys

For most metals, if they are stretched a little, they can spring back to their old shape. However, when stretched further or bent, they stay that way. If a part made from a **shape memory alloy (SMA)** is bent out of shape, when it is heated to a certain temperature it will return to the original shape of the part. This temperature is known as the **transition temperature**. This cycle of bending and being straightened can be repeated many times. SMAs can be formed into almost any shape, from springs to flat plates.

The most common SMA is an alloy of the metals nickel and titanium, which has a transition temperature of 70°C. This means that it can return to its original shape by, for example, being put in hot water.

Uses of shape memory alloys
- Triggers to start the sprinklers in fire alarm systems.
- Controllers for hot water valves in showers or coffee machines.
- Artificial muscles in robot hands.
- Spectacle frames (Figure **A**).

Piezoelectric materials

Piezoelectric materials do not conduct electric current. However, if their shape is changed quickly, they produce an electrical voltage for a moment. This also works in reverse: if a voltage is put across the material, it makes a tiny change in shape.

Piezoelectric materials have already found many uses, such as contact sensors and as motors to move camera lenses. They are also an essential part of many microphones and the speakers in miniature headphones.

A Spectacle frames made from shape memory alloy

Quantum tunnelling composite

Quantum tunnelling composite (QTC) is a flexible polymer. It contains tiny particles of metal. Normally, electricity cannot pass through it. However, when squeezed, electricity can pass through it. QTC can be used to make switches like those used for the buttons on mobile phones or pressure sensors.

Colour change materials

Thermochromic materials change colour as the temperature changes. They are available in lots of different forms – for example, as solid plastic, as dye for clothes or even as paint. They already have several uses:

- Plastic strips that use colour changes to indicate temperature (Figure **B**).
- Test strips on the side of batteries. These heat a resistor printed under the thermochromic film. The heat from the resistor causes the film to change colour.
- Packaging materials for food. They change colour to show you when the product they contain is cooked to the right temperature.
- Colour indicators on cups, to show whether the contents are hot.

Photochromic materials change colour according to different lighting conditions. They are used for products ranging from nail varnish, security markers that can only be seen in ultraviolet light, jewellery and mobile phone cases.

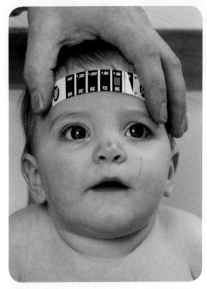

B Contact thermometer made from thermochromic sheet

Activities

1 Choose one type of smart material. Design an A4 advertisement that would encourage designers to use it.

2 Choose a type of smart material different from that in Activity 1. Write a short report explaining which other materials it could be used to replace, explaining its potential advantages and drawbacks.

Case study

Smart skis

When skiing fast on rough terrain, skis tend to vibrate. This makes skiing harder to control and normally slows the skier down.

Smart skis contain piezoelectric sensors. The vibration changes the size of the piezoelectric material, sending electrical signals to a control system. This, in turn, sends electrical signals to the piezoelectric material, which cause it to change shape. The shape change alters the stiffness of the ski and reduces the vibration. This gives the skier more control, a smoother ride and allows them to go faster.

Can you think of any other sports-related applications where piezoelectric sensors could be used to improve performance?

Summary

- Smart materials can change one or more of their properties in response to changes in their environment.

- There are smart materials that can change shape with temperature, change electrical resistance with pressure or change colour with temperature or light levels.

8.1

Materials removal

Objectives

- Be able to list a range of tools that can be used to remove material.

Key terms

Wasting: using tools to remove the unwanted material from a product.

Different tools are used to carry out different tasks. One of the most common types of task is **wasting**. This is where unwanted material is cut away from the product. A wide variety of tools are used to carry out wasting processes in school workshops. Only a small number of these tools are covered on these pages, but they should give you an indication of the range of tools that is available.

Tools for sawing and separating material

One of the most common types of tool used for wasting is the saw (see the table below).

Different types of saw have different sizes of teeth, so that they can cut different types of material. The harder the material, the smaller the saw's teeth. The teeth are normally bent outwards slightly. This stops the blade jamming, but means that the width of the cut is slightly bigger than the width of the saw blade.

Guillotines and shears can be used to cut metal sheet. These use a large force to press a blade made of tool steel through the material.

Adhesive vinyl (sticky-back plastic) can be cut using a computer-controlled knife in a vinyl cutter.

Parts can also be cut by melting along the line to be cut. This is done using a heated wire or a laser cutter for thermoplastic materials, and an oxyacetylene torch for ferrous metals.

Tool	Material	Process
Tenon saw	Wood	Used to make straight cuts. The blade is stiffened to help keep the cut straight
Coping saw	Wood and plastics	Used to make curved cuts. There is a thin blade, so it is easy to change the direction of cutting. The blade is held in tension in a steel frame. A scroll saw is similar to a coping saw, but with a motor to move the saw blade
Hacksaw	Metal and plastics	Used to make straight cuts. The teeth are much finer than those in a saw used to cut wood. The blade is normally held in a steel frame and put under tension by a screw at the front of the hacksaw
Junior hacksaw	Metal and plastics	Used to cut thin pieces of metal or plastic. This is a smaller version of the hacksaw. The blade is held in tension in a steel frame

Tools for materials removal

Filing is done using files. A file is a hand tool. It has hundreds of small teeth to cut away at a material. Rough cut files are used to remove material and fine cut files are used for finishing, as they give a smooth surface. Most files are meant for metal and plastic. There are a lot of different shapes of file (Figure **A**).

Drilling is used to make holes in wood, plastic or metal. If the part being drilled can be moved, a pillar drill is usually used (Figure **B**). If the part cannot be moved, a battery powered drill is used. The tool put into the drill and used to make the hole is called a 'drill bit'. Different types of drill bits are used for different materials.

Sanding involves using an abrasive tool to wear away the surface of a wooden product. It can be carried out by hand, or by using machines such as a belt sander or disc sander (Figure **C**). Sanding is used to remove unwanted material or to make a surface smoother.

Turning is where round objects are made using a lathe. There are different types of lathe for wood (Figure **D**) and metal. The piece of material is held by the machine and rotated at high speed. The tool is pressed into the work piece to remove material.

Milling machines are used to remove material one thin layer at a time (Figure **E**). For metal, they produce a flat surface with a good finish. Many schools have small computer numerical controlled (CNC) milling machines, which are sometimes called CNC routers. These can also be used with wood and some plastics.

Grinding machines use spinning wheels made of abrasive materials. Each time they touch the part they remove less material than milling. They can be used to accurately shape metal parts and to produce a very smooth, mirror-like finish.

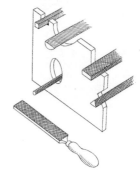

A Different shapes of file are used for different shapes on the products

B Pillar drill (drill guard removed for clarity)

C Disc sander

D Wood being turned on a lathe

E Manual milling machine

Activities

1 Sketch some of the different tools and equipment available in your school. Label any important features and include an explanation of how it should be used.

2 Examine a wooden chair or table. Identify which tools could have been used to make it.

Summary

● Tools and equipment used for cutting include saws, guillotines, hot wire cutters and laser cutters.

● Tools and equipment used for materials removal include files, drills for making holes, lathes for turning, mills and grinders.

Objectives

- Be able to describe processes used to form wood, metal and thermoplastic materials.

Forming involves using force to change the size or shape of a piece of material. The piece of material to be formed is normally a flat sheet. Things that you might form include chair seats, baths and the bodywork for cars.

There are cold and hot methods of forming. Cold forming methods are used for metal sheet and thin strips of wood, as their shape can be changed without breaking the material. Hot forming methods use heat to soften the material. This makes it easier to form. Hot forming is widely used for thermoplastic materials.

Cold forming

Laminating wood

Laminating is used to make shapes from wood that cannot be cut from one piece of material. These are used for items ranging from the seats for chairs to curved roof beams.

A A laminated chair

The first step in laminating is to take several very thin strips of wood and put glue between them. While the glue is still wet, these strips are squeezed into the shape needed between two formers (Figure B). Once the glue dries, the laminated wood stays in the shape needed.

Former

Pressure

B Laminating wood

Press forming metals

In press forming, a sheet of metal is placed between two formers. These formers are normally made from very hard tool steel. The press then pushes down on the formers with great force – this might be hundreds of tonnes in an industrial press. This huge force deforms the metal into the shape needed. In industry, automatic presses are used to form shapes such as car door panels.

C Example of a press formed part: car body panels

Hot forming methods

Strip bending thermoplastics

Strip bending is often done in schools with acrylic sheet. It involves heating a sheet of plastic along a line, using a heating element. As the plastic heats it softens, allowing it to be bent. As it cools it will keep its shape.

It is a good idea to make a wooden former to ensure accurate **bending** (Figure **D**). It is important to allow the plastic to cool slightly before it is removed from the former – otherwise, if it is still hot it might sag, causing its shape to change.

Vacuum forming thermoplastics

Vacuum forming is used to make many different products from thermoplastic sheets. These include packaging, helmets, masks and baths. The sheet is heated to make it flexible, formed over a mould and then cooled to become hard again (Figure **F**).

The moulds used in school are often made from wood or MDF. The corners of the mould should be rounded. The sides of the mould must have a slope to allow the plastic product to be lifted off (Figure **E**). This slope is called the **draft angle**. It should be between 5 and 10 degrees. If it had square edges and no angle, the plastic product might stick into the mould.

Vacuum forming can only be used to make shapes of simple profiles, as any overlaps would cause the plastic to stick on to the mould.

Acrylic book stand

Wooden mould to ensure accurate bending of book stand

D Strip bending

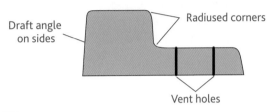

Draft angle on sides

Radiused corners

Vent holes

E Vacuum forming mould

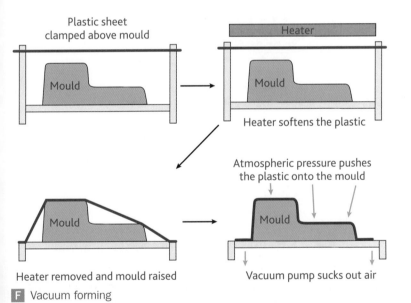

Plastic sheet clamped above mould

Mould

Heater

Mould

Heater softens the plastic

Atmospheric pressure pushes the plastic onto the mould

Mould

Mould

Heater removed and mould raised

Vacuum pump sucks out air

F Vacuum forming

Key terms

Laminating: gluing thin strips of material together to make a thicker piece.

Bending: forming an angle or curve in a single piece of material.

Vacuum forming: forming a thermoplastic sheet over a mould, using heat and a vacuum.

Summary

● Laminating involves sticking thin sheets of wood together and allowing the glue to dry while they are squeezed between two formers.

● Metal can be cold formed using direct force from a press to change its shape.

● Strip bending and vacuum forming use heat to soften and form thermoplastic material.

Activities

1 Find an example of a product made using each of the four forming processes described on these two pages. Explain what features of the product show that it was made by that process.

2 Design a vacuum forming mould that could be used to make a fishpond for a garden. Explain each of the features of your mould and why it is suitable for use with the vacuum forming process.

Objectives

- Be able to describe processes used to shape metal, plastic and composite materials.

Shaping processes normally involve making parts by using liquid materials. These processes are used for metals, plastics and composite materials.

Injection moulding of thermoplastics

Injection moulding is used to make a wide range of plastic products, ranging from bowls, buckets, model construction kits to chairs and toys. The process is very fast and complicated shapes can be made from both thermoplastic and thermosetting materials.

Figure **A** explains how injection moulding is carried out. It is normally easy to identify an injection-moulded part, as the **sprue point** where the plastic was injected is often visible (Figure **B**). There may also be a **split line** visible if the sides of the mould did not fit together perfectly.

1 An expensive mould is made, normally from steel.

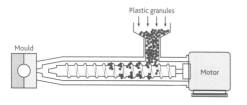

2 Granules of plastic are placed in the hopper.

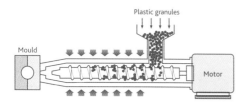

3 Heaters melt the plastic into a thick liquid which is pushed towards the mould by an Archimedean screw.

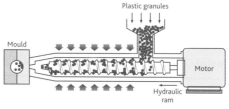

4 A hydraulic ram forces the plastic under huge pressure in to the mould.

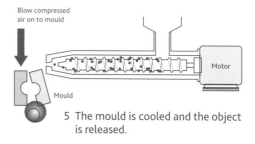

5 The mould is cooled and the object is released.

A The injection-moulding process

@ Links

See **7.5 Ceramics and composites** for more information on composites.

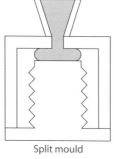

Split mould

Finished bottle top before sprue is cut off

B Injection moulding of a bottle top, showing the sprue still attached

Casting

Casting is used to make 3-D shapes from metal. It is normally much cheaper and quicker to cast a complicated shape than to machine it from a solid piece of metal. Cast products include engine blocks, camera bodies and door knobs.

C   An MDF mould for pewter casting

During the process of casting, the metal is heated above its melting point so that it becomes a liquid. The liquid is poured into a **mould** and takes the shape of the hollow area in the mould. This hollow area is called the cavity. Once the material cools and becomes solid again, it is removed from the mould.

Most metals can be cast. Iron and aluminium can both be cast into moulds made by pressing shapes into sand. Zinc, which has a lower melting point, is often cast in metal moulds. Pewter, which is often used in schools, can even be cast into moulds made from MDF.

Key terms

Injection moulding: the process of making plastic parts by forcing liquid plastic into a mould and allowing it to solidify.

Casting: the process of making parts by pouring liquid metal into a mould and allowing it to solidify.

Mould: a former used to shape a part.

Moulding composite parts

Fibreglass is a composite material that is tougher than steel. It is made from flexible glass fibres and a thermosetting resin, which is hard and brittle when it sets. Fibreglass products are easy to identify as they are smooth on one side and rough on the other. Fibreglass has many uses including canoes, car bodies and children's playground equipment.

Female mould

Rough surface

Smooth surface

Male mould

Rough surface

Smooth surface

D   Forming of fibreglass

Fibreglass products are usually shaped by hand, using a one-piece mould. It is important that the mould is made the correct way to make sure that the smooth side faces the direction you want (Figure **D**). The mould is covered in a layer of resin and then a layer of glass fibre. A layer of resin is applied, to soak into the glass fibre, then another layer of glass fibre. This is repeated until the thickness needed is achieved, finishing with a final layer of resin. It is important to make sure that the layers of glass fibre are pushed into any corners. The part is then allowed to stand for up to 24 hours for the resin to go hard.

Summary

- Casting involves heating a metal above its melting point and pouring it into a mould.

- Injection moulding involves forcing liquid plastic into a mould. Both casting and injection moulding can be used to make complicated 3-D shapes.

- Fibreglass is normally shaped in small quantities by hand, using a mould.

Activities

1   Find a plastic product that has been made by injection moulding. See if you can identify the sprue point.

2   Identify three products made using each of the three shaping processes described on these two pages. Explain why these products are made using that process.

Joining wood

Objectives

- Be able to describe different types of wood joints.

Traditional wood joints

There are several different ways of joining two pieces of wood together. When choosing which joint to use, you need to think about how strong it needs to be and how much effort is needed to make it. Many of these joints use, or can be reinforced by, glue. The most common adhesive that is used with wood in schools is **polyvinyl acetate (PVA)**. This soaks into the surfaces of the wood, creating a strong bond when it dries.

Butt joint The simplest form of joint and also the weakest. It is normally glued together, but might be reinforced by nails or screws.		**Lap joint** This is a little stronger than a butt joint, as there is a bigger surface area for gluing. These are used in low-cost furniture, for example to attach the backs to cupboards.	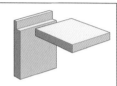
Halving joint This is made by removing half the material thickness from the two parts to be joined. They are normally glued together. It is stronger than a butt joint, as there is more contact area between the wood.	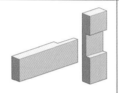	**Dowel joint** This is stronger than a butt joint, but not as strong as a mortise and tenon joint. It is easy to make as it only needs holes to be drilled for the dowels. It is often reinforced with glue.	
Dovetail joint This is the strongest joint for box constructions, due to the shape of the 'tails' and 'pins'. It also looks decorative. However, it is difficult to cut by hand. It is used in high-quality boxes, drawers and furniture.	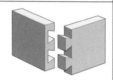	**Mortise and tenon joint** A strong joint used in high-quality furniture. It can be reinforced by using glue.	
Housing joint This is slightly stronger than a butt joint as there is more surface area for gluing. The slot is often cut with a power tool, such as a router. This is often used to put shelving together.		**Biscuit joint** This joint is quick to make. The slots are cut with an electric cutter and the wooden biscuits are glued into place. This is often used to attach flat surfaces together, such as kitchen worktops.	

Alternative joining methods

Nails and screws

Nails are low cost, quick and easy to use – you only need a hammer! They make a 'semi-permanent' joint. This means that the parts they join could be taken apart at a later date, although it might be quite difficult to do so. Most nails are made from steel. You can get nails made from copper or aluminium for specific purposes, but these are more expensive.

A Round wire nails

Screws can be used to join a wide variety of different materials, not just wood. Compared to nails, it takes slightly longer to make a joint with screws. However, the joint is often a bit stronger. It is also easier to take the joint apart later, if needed. Both nails and screws can be bought in a wide range of different sizes.

Knock-down fittings

Knock-down fittings were developed for use in flat-pack furniture. 'Flat pack' means that when you buy the furniture you get all the parts cut to the right size and shape, but you have to put them together yourself. Typical items of flat-pack furniture include kitchen cupboards, shelves and tables. The advantage of being flat pack is that they take up little space and are easier to store and move around than a finished piece of furniture.

B A selection of specialised screws. Note the variety of head shapes and threads

Key terms

Knock-down fittings: fittings used to join together flat-pack furniture.

Knock-down half block This screws directly into two parts to hold them at a right angle.	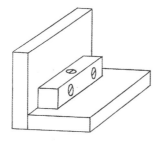
Knock-down full block This attaches separately to two parts, which can then be joined with a single screw between the parts of the block.	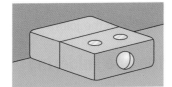 Block connector
Corner plate This can be used to reinforce a butt joint. It can also allow a table leg to be attached, using the central hole with a special threaded fitting and a wing nut.	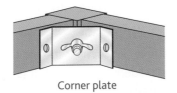 Corner plate

Activity

1 Examine an item of wooden furniture in the workshop – for example, a cupboard or a shelf unit.

   a Identify the types of joint used and any fixings – nails, screws and knock-down fittings.

   b Explain why the designer chose these instead of other types of joint or fixings.

Summary

● There are a wide variety of different types of wood joints, including butt, dowel and dovetail joints.

● The choice of joint depends on the strength needed and the effort required to make it.

● Alternative joining methods for wood include nails, screws and knock-down fittings.

Joining metals and plastics

Objectives

- Be able to describe different ways of joining metals and plastics.
- Understand the purposes for which these are used.

Welding

Most joints involve two pieces of material that need to be attached to each other. **Welding** involves heating the area where the pieces of material touch each other, until they melt and run together. As there is often a small gap between the parts being joined, you often need to use a filler material.

Welding can only be used to join materials that are similar to each other. For example, you could join two pieces of aluminium together, but you could not weld a piece of aluminium to a piece of steel or a piece of metal to a piece of plastic. Welding forms a permanent joint. This means that either the part or the joint would have to be destroyed to break it. Welds are the strongest type of joint that can be formed.

Welding processes for metals include **manual metal arc (MMA)** welding and **metal inert gas (MIG)** welding. These use the heat from an electric arc to melt the metal and form the joint.

Welding processes for thermoplastics include hot plate welding, hot wire welding and friction welding. These involve heating the surfaces to be joined, either directly or by friction, and then pushing the parts together to form the joint.

A Manual welding

Brazing and soldering

Brazing uses heat to join metals. However, unlike welding, the parts being joined are not melted. Only the brazing material that is added is melted and this flows into the joint. Brazed joints are strong, but not as strong as welded joints.

Soldering is very similar to brazing. It is likely that you will use manual soldering at school. It is mainly used to attach components to circuit boards.

B Manual soldering

Riveting

Riveted joints are used in applications such as attaching the skin to aircraft. The rivets are inserted into holes in overlapping pieces of material and the ends are made bigger to hold the materials in place (Figure **C**). Rivets are mainly used to join metals.

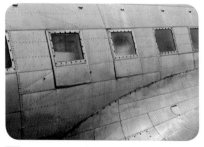
C Metal sheets joined by rivets to form the skin of an aeroplane

Adhesives

There are a wide range of **adhesives** that can be used to bond parts together (see the table below). Some of these can be used to join together different types of materials. Most adhesive joints are weaker than the parts being joined. It is important that the surfaces being joined are free from dirt and grease.

Name	What it is used for	Any other information
Polyvinyl acetate (PVA)	To join pieces of wood together	A white, water-based adhesive. It soaks into the surface, forming a strong bond on setting
Epoxy resin	Can join most materials together, even different ones (e.g. plastic to metal)	Stored as two separate chemicals. The two parts have to be mixed together immediately before it is used
Contact adhesive	Joining plastic parts, bonding plastic to wood	The solvent fumes can be dangerous. Both surfaces must be coated with this and allowed to become touch dry before being forced together
Acrylic cement	Joining pieces of plastic	This dissolves the two surfaces to be joined, allowing them to mix together and set, similar to a weld
Hot melt glue	To join plastic and/or wood together	Pale or opaque glue sticks that are melted in a hot glue gun. The glue can be very hot and can cause nasty burns if you are not careful

Threaded fastenings

Threaded fastenings include nuts, bolts and screws. These are available in a wide range of materials, including steel, brass and different thermoplastics. This type of fixing can be used to make joints in most materials, including joining two different materials.

Threaded fastenings form temporary joints. This means that they can be taken apart and put back together again. This is useful if you need to repair a product. These joints are not as strong as permanent methods of joining and they can sometimes become loose over time.

Activity

1 Carry out a product analysis of a mobile phone, a remote control or a similar small product that you can take apart. Sketch the product and identify the different joining processes that were used to make it.

Key terms

Welding: a method where a joint is created by melting the contact areas of the parts being joined.

Adhesives: compounds used to chemically bond items.

Threaded fastenings: mechanical parts such as screws, nuts and bolts.

Links

See **8.4 Joining wood** for more information on threaded fastenings.

See **1.4 Product analysis** for more information on product analysis.

Summary

- Metals can be permanently joined by welding, brazing, soldering, riveting or using adhesives.
- Plastics can be permanently joined by welding or using adhesives.
- The most common temporary joining method is to use threaded fasteners, such as screws, nuts and bolts. These can be used to join different materials to each other.

Finishing

Key terms

Surface finishing: modifying the surface of the part in a useful way.

Polishing: a physical process where the surface is rubbed or buffed to make it smoother.

Painting: applying a liquid that dries to form a coating on the part.

Plating: depositing a layer of metal using electrolysis.

What is finishing?

Surface finishing involves changing the surface of a product. It is carried out to improve the properties of the product. The finish might protect it from corrosion or make it more resistant to scratches, or it might improve its appearance.

The finish might be an extra layer, a coating, a decoration or an extra process that is used on the surface of the product. The type of finish depends upon the type of material and the purpose of the product.

Finishes for plastic products

Finishing is not normally used with plastic products. They are already coloured and the manufacturing processes used to make them give a high-quality surface. The only process that might be used is **polishing** with a buffing machine. This can be used to remove small scratches and marks.

Finishes for wood

Type of finish	How it is applied	Properties of the finish	Environmental impact
Wax	Rubbing into the surface with a cloth. Once dried, it needs to be buffed	Adds a shine to the surface and some protection	No pollution
Oil extracted from wood	Rubbing into the surface with a cloth	Adds a shine to the surface and some protection. Can make the grain stand out to improve the appearance	No pollution
Stain	Using a cloth or a brush	Changes the colour of the wood, but provides little protection. Often used with a sealer to add protection after it has dried.	Pollutant when liquid
Sealer	Using a brush	Adds a shine to the surface and good protection.	Pollutant when liquid
Varnish	Using a brush or a spray	Adds a shine to the surface and good protection. Can also be used to add colour	Pollutant when liquid
Paint	Using a brush or a spray	Adds colour to the wood and good protection. Can hide the natural appearance of the wood	Pollutant when liquid

A Teak furniture – this is often protected using teak oil

Finishes for metal

Painting

The cheapest and easiest way of coating a metal item is often **painting**. This adds colour and can give good protection against corrosion and rusting. Paints can be applied by brushing, spraying or dipping.

Dip coating

The first step in dip coating is to heat the metal part to about 200°C. It is then dipped in a fluidised bath of plastic powder. The plastic sticks to the hot metal and is left to cool. This gives a strong coating that provides excellent resistance to corrosion and a high-quality appearance.

Galvanising is similar to dip coating. It involves dipping the metal in liquid metal zinc (rather than the plastic powder). It gives excellent corrosion resistance but the appearance can be poor. Metal dustbins are often made from galvanised steel sheets. Galvanised parts are sometimes painted afterwards.

Plating

Plating uses a process called **electrolysis** to coat a thin layer on the surface of a metal. This means that the part to be coated is put in a tank containing a chemical solution and an anode. An electric current is passed through the part. This causes a thin layer of metal to slowly build up on it (Figure **C**).

Plating is used to apply coatings of chromium, nickel, zinc, copper or tin. Nickel and zinc coatings provide good corrosion resistance. Chrome plating is used to give a good appearance on taps (Figure **D**) and the handlebars of bicycles and motorbikes. Parts that are chrome plated are normally nickel plated first, to improve their corrosion resistance.

B Number 10 Downing Street, the world's most famous painted metal door!

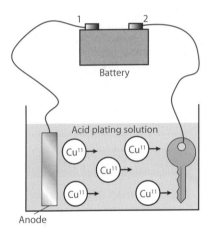

C Electroplating a key

D Chrome-plated bathroom taps

Activity

1 Identify at least five different products that have some form of coating. Explain the benefit that the coating or surface finish brings to each of these products.

Summary

- Surface finishing involves changing the surface of a product. It might be used to protect the product or to improve its appearance.

- Finishing is not normally used for plastic products, although they might be buffed if necessary.

- Wood products can use a variety of finishes, including wax, oil, stain, sealer, varnish and paint.

- Finishing techniques used with metal products include painting, dip coating and plating.

Health and safety

Objectives

- Take responsibility for safe working.
- Be able to give some general guidance for safe working.

Key terms

Hazards: things that could cause injury or harm.

Personal protective equipment (PPE): equipment worn or used in the workshop to protect against hazards.

Activity

1 Draw a map of your school workshop. Identify the entrance and exit points, emergency stop buttons, safety equipment and any possible hazards.

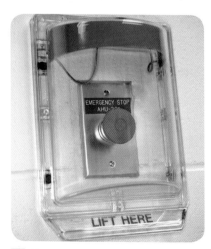

A Emergency stop button

School workshops can be very dangerous places. Accidents in the school workshop can cause broken bones, the loss of fingers or limbs, or worse.

These accidents are often caused by carelessness. You must always concentrate on working safely. This is vital for both your own safety and that of others using the workshop. It is *your* responsibility to behave in a mature and responsible manner, maintain a safe environment and use safe working practices.

Common causes of accidents in school workshops

- Carelessness or loss of concentration.
- Not following safety rules.
- Lack of training in using the equipment.
- Unguarded or badly maintained equipment.
- Tiredness and fatigue.
- Slippery floors and messy work areas.
- Poor behaviour – not working sensibly.
- Trying to lift items that are too heavy.

Reducing the risks of injury

The first step towards working safely is to identify the **hazards** in the workshop. Hazards are the things that could cause injury or harm. If you know what the risks are, you can take actions to reduce them.

The following points are some of the typical actions to reduce risks. There will also be actions unique to the workshop that you work in and for the equipment that you use.

- Make sure that you know how to carry out any activities that you need to do safely. These may include, for example, using machine guards and following safe working procedures.
- Be aware of what to do in case of an emergency. For example: check out the fire procedures and be aware of the locations of the different types of fire extinguishers; know where the emergency stop buttons are; find out what to do if there is an accident.
- Do not use any equipment that is damaged or incomplete. For example: don't use electrical equipment if its wires are frayed or damaged; don't use a machine if its guard is cracked. Tell whoever is responsible for supervising the practical work about the problem immediately.
- When using sharp tools point them away from your body and keep both hands behind the cutting edge.

- Don't lift heavy items by yourself. Ask for help if needed.
- Always clean away any mess that you make. This includes returning any tools and materials to the correct places. Keep areas between benches and machinery clear to avoid tripping.
- Carry tools and materials in a safe manner with sharp points and edges pointing away from people.
- Only use chemicals after reading all the instructions. Ask your teacher about disposal of chemical waste.

Personal protective equipment

Sometimes, the risk of injury can be reduced by wearing **personal protective equipment (PPE)**. Each piece of PPE is designed to protect against specific hazards (see table below).

Operation	PPE	Hazard	Safety symbol
Drilling, sanding, welding	Goggles, welding visor	Dust, swarf or sparks might get into eyes	Eye protection must be worn
General workshop activities	Apron	Clothing could be caught by tools or in machinery. Dust and chemicals can get spilt on to clothing	Protective clothing must be worn in this area
Handling hot/ sharp materials	Heat-proof gloves, leather apron (plus leather leggings and full face mask if aluminium casting)	Burned hands/ fingers when working with hot materials. Cut hands when carrying sheet materials such as steel or glass	
Using machinery	Ear defenders	Damaged hearing from repetitive or continuous loud noise	
Sanding, applying a finish, using adhesives	Face mask, latex gloves	Lung damage from inhaled dust or fumes	Wear face mask
Carrying or installing equipment	Stout shoes with toe protection	Damaged or crushed toes and feet caused by falling materials or equipment	Foot protection must be worn

Activity

2 A group of Year 6 pupils are going to be visiting the workshop in your school. They will be making a simple wooden product using hand tools. You have been asked to tell them about safety in the workshop. Prepare a short presentation, which explains the safety rules and any other safety information that they need to know.

Summary

- Everyone in a school workshop should act and work in a way that maintains a safe and healthy working environment.
- Accidents in a school workshop can be prevented by identifying hazards and reducing the risks.
- PPE can be used to reduce some of the risks in a workshop.

Objectives

- Be able to explain what is meant by the term 'rapid prototyping'.
- Understand why rapid prototyping is used as part of the design process.

Key terms

CAD CAM: a process used to produce products using computer-controlled output devices.

Rapid prototyping: a CAD CAM system used to make accurate 3-D models.

Stereolithography (STL) file: a file format used for rapid prototyping and CAM.

CAD

Computer-aided design (CAD) is a way of generating and developing design ideas using a computer. There are many different types of CAD software used in schools.

CAD CAM

CAD can also be used as part of a system to make things. This process is called **CAD CAM**. Computer-aided manufacture (CAM) is used to manufacture items designed using CAD. In CAD CAM, information produced on a computer, usually in the form of a drawing, is sent to a machine that then makes the product.

There is a wide variety of CAM equipment available in schools, such as laser cutters, milling machines, routers and, more recently, **rapid prototyping** systems (also known as 3-D printers).

Rapid prototyping

Designers need to evaluate design ideas before manufacturing begins. This is an important part of the design process. If problems are not identified at an early stage, mistakes in a mass-produced design could be expensive to correct. Traditionally, designers would send detailed drawings to model makers, who would then produce accurate representations of the design and return them to the designer to evaluate. This process took a lot of time and often required a large number of models to be made.

The introduction and development of rapid prototyping technology means that designers can now produce these models themselves, which saves time and money. Rapid prototyping makes models quickly and very accurately. It has been an important factor in saving time when developing new products to be sold. As this technology has become more established, and costs have decreased, rapid prototyping has gradually been introduced into schools.

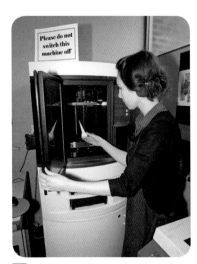

A A student using a rapid prototyping machine in school

Links

See **4.5 CAD** for more information on CAD.

See **5.3 CAM and CIM** for more information on CAM.

Stereolithography

In rapid prototyping systems, CAD drawings are exported in a **stereolithography (STL) file** format to the rapid prototyping machine. STL software reads the information from the drawing and converts it into horizontal 'layers'. The rapid prototyping machine then builds the model layer by layer using the information from the drawing.

Rapid prototyping systems

The most common forms of rapid prototyping in schools are:

- **Inkjet printing**: a print head moves across a table containing a thin layer of powder depositing a liquid binder to form one layer of the model. The table then moves downwards, a new layer of powder is applied and the print head moves across the powder again, forming the second layer of the model. This process continues until the whole model is created. Each layer is bound together by the liquid binder. Once the liquid has set, the model can be removed from the machine.
- **Fused deposition modelling**: this system works by using a print head to deposit a molten polymer on to a platform, which is gradually lowered, layer by layer. Each layer is fused to the previous layer. Once the finished model has cooled down, it can be removed from the machine.
- **Digital light processor (DLP)**: a liquid polymer is exposed to light from a DLP projector. The exposed liquid polymer hardens, the build plate moves down and the process is repeated until the model is built. The excess polymer is then drained from the tank and the model removed.

B Modelling a design using CAD

C The finished product from the design shown in Figure **B**

In schools, rapid prototyping allows complex forms to be made without the need for traditional 'making' skills. Rapid prototyping requires a high level of skill in CAD, and the prototypes made in this process cannot be altered easily without being completely remade.

Link

Find out more information on rapid prototyping and 3-D printing at:

www.dimensionprinting.com

www.zcorp.com

Summary

- Computers can be used to design and make products.
- CAD CAM is a process used to produce high-quality products using computer-controlled output devices.
- Rapid prototyping is a CAD CAM process used by designers to make very accurate models.

Activity

1 Why do you think designers make models of their designs before they are manufactured commercially?

Objectives

- Be able to understand how and why products are designed for volume production.
- Understand why jigs and fixtures are used to ensure accuracy and consistency.

Key terms

Production line: a set of operations designed to produce a product as efficiently as possible.

Jig: a device used to hold or position a work piece for machining in order to achieve a consistent and repeatable end result.

Fixture: a device, similar to a jig, used to hold or position a work piece for machining that is attached to a machine.

 Links

See **5.2 Scale of production** for more information on scales of production.

See **8.3 Shaping** for more information on injection moulding.

See **8.8 CAD CAM and rapid prototyping** for more information on rapid prototyping.

Design considerations

Products designed for commercial production will often be designed and produced in a different way to bespoke ('one-off') items, or a product made in school.

The main objective for commercial manufacturers is to make as much profit as possible. This means that they need to produce each product as cheaply as possible. The best way to do this is by designing products that can be manufactured using batch production or mass production techniques.

However, cost is not the only requirement when designing a product for volume production. Designers also need to think about how the consumer will use the product. For example, some products may need regular maintenance or worn out parts may need to be replaced.

It is also important that the product is designed to be manufactured efficiently. This might be by using a **production line**. Mass-produced products are often made using automated production lines and CAM processes.

To manufacture products in quantity, manufacturers often need to use specialised machinery and equipment, and even different processes to those used to make a prototype. They will carefully consider the choice of material used to make the product because some materials are more suited to volume production than others.

Materials used to model a prototype are often different to materials used for the final product. Injection moulded components are often developed using rapid prototyping technology, but the material used for this process is far less durable than the plastics used for the real product.

A An automated production line

Jigs and fixtures

An important aspect of volume production is the use of **jigs** and **fixtures**. A jig is a work holding, or positioning device, which allows parts to be made quickly and consistently. It is an important feature of a mass-produced product. A fixture is similar to a jig, but is normally fixed to the bed of a machine. Jigs and fixtures are used to ensure that the same parts are produced identically each time.

For example, if three holes have to be drilled accurately in 50 pieces of steel, a jig would be designed to hold the work piece in the same position each time and guide the drill bit so that every hole was in the right place.

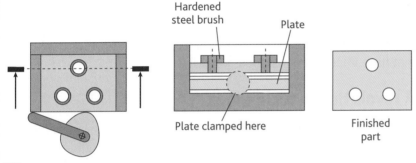

B A jig used to drill holes in a batch of steel plates

Without jigs and fixtures, you would need a skilled worker to produce the part from a drawing each time the product was made. If this part were to be produced on a much larger scale, a computer-controlled multi-headed drill would be used to drill the three holes at the same time (Figure **C**).

C A multi-headed, computer-controlled drilling machine

Activities

1 Some products are assembled by hand on a production line. Select a product that has been made this way and produce a flow chart showing how you think it was assembled.

2 Think of a product you have made. Using notes and sketches, explain how you think it could be modified for commercial production.

Summary

● When making products in quantity designers must consider how the product will be manufactured.

● Production lines are used to make products efficiently.

● Jigs and fixtures are used to ensure consistency when making components in quantity.

Manufacturing in industry

Objectives

- Understand the impact of modern manufacturing on the way we live today.
- Understand how manufacturing has been shaped by the introduction of new technologies.

 Link

See **8.8 CAD CAM and rapid prototyping** for more information on CAD CAM.

Taking things for granted...

We often take for granted the things we use every day, without thinking about how these products have been designed and made. The way we make things has changed enormously over the past century. Developments in engineering, materials and technology have shaped the way we live our lives today. An important part of this is the way products are manufactured.

Today, new manufacturing processes, new materials and the widespread introduction of CAD and CAM mean that a huge range of products are available to everyone, at an affordable price. These developments have allowed designers to develop innovative products that have improved our standard of living.

A revolution in technology

Before the industrial revolution, materials were often limited to those that could be produced by hand from local resources. Here are just some of the events that shaped the early development of modern manufacturing:

- The industrial revolution saw a rapid growth in global trade, which led to raw materials becoming available from around the world.
- The introduction and development of the steam engine, together with iron and steel production, led to the introduction of mass-produced goods.
- The development of synthetic plastics brought about another revolution in manufacturing technology.

The development of machines

Most manufacturing processes are carried out using machines. Until the 1950s, all machining processes were carried out on manual machines. These machines are still produced and used as they are suitable for many applications (Figure **A**).

A A manually operated milling machine

However, the first electronically-controlled machine was introduced in 1952. These machines could produce difficult shapes repeatedly, faster and more accurately than skilled human operators. Only five years later, the earliest computer numerically controlled (CNC) machines began to replace some of these electronic machines (Figure **B**).

B A computer-controlled milling machine

Practically every manufacturing company now uses some form of CNC machine. For some companies that manufacture large quantities of parts, every machine may be CNC. CNC machines have led to greater accuracy of machining, the ability to make more parts and reduced product costs.

CAD

The development of CAD has reduced the constraints on the designer and brought about the 'blobject': a product with a more free-flowing shape, which could not be easily designed by hand. Products designed this way can then be manufactured using injection moulded plastic and cast or pressed metal. Examples include the Apple iMac computer and the Audi A1.

C An Apple iMac computer

D An Audi A1

Computers in manufacturing

Computers are also vital to the process of modern manufacturing. Many modern factories are now fully automated. Computers are often integrated into every function, from placing the initial order to organising delivery of the finished product to the customer.

In continuous production, the production line never stops, running 24 hours a day, seven days a week. Continuous production relies on high levels of automation, involving CNC machinery and robot assembly lines.

Manufacturing: the future

Traditionally, manufacturers have used materials from non-renewable resources and processes powered by fossil fuels. There is a growing awareness that these resources will eventually run out. Because of this, the challenge for the future is to develop products and processes that have a lower impact on the environment. The target is **sustainable manufacturing**.

E An automated production line

 Link

See **2.1 Environmental concerns** for more information on sustainability.

Key terms

Blobject: a product designed using CAD or CAM to reduce designing constraints.

Sustainable manufacturing: products that are made using processes that are non-polluting and that conserve energy and natural resources.

Summary

● Manufacturing has changed because of developments in manufacturing processes, materials and technology.

● The use of computers for CAD and CAM has revolutionised modern production processes.

● The need for sustainable design will influence future developments in manufacturing.

Activities

1 Consider a product that you have recently bought.

   a Could fewer raw materials have been used in its manufacture?

   b Can the materials be easily recycled or are they biodegradable?

   c Could the amount of packaging have been reduced?

2 Describe what you think is meant by the term 'sustainable manufacturing'.

Integrating systems

Many products are designed to do something. This means that they have functions built-in to make them more useful or flexible. Advances in technology have given designers the flexibility to produce increasingly more sophisticated and functional products by **integrating systems** into their designs.

Often, the newer the product, the more things it can do. Mobile phones do much more than make and receive calls. Modern 'smartphones' are, in fact, small computers. A mobile phone is used to send a text message far more often than it is used to make a call. Some smartphones can even be used for satellite navigation or 'satnav'. The introduction of 'apps', or applications, for mobile phones means that making telephone calls is now only a small part of how a phone is used.

Integrating systems into products enhances their **functionality**.

Mobile phone comparison

Function	iPhone 4 (2010)	Nokia 5110 (1998)
Size:	115.2 x 58.6 x 9.3 mm	132 x 47.5 x 31 mm
Weight	137 g	170 g
Display type	Touchscreen-16M colours	single colour – black
Display size	640 x 960 pixels, 3.5 inches	5 lines of text
Messaging	SMS and email	SMS
3G	yes	no
WLAN	yes	no
Bluetooth	yes	no
USB	yes	no
Camera	5MP, autofocus, LED flash	no
Video	yes	no
GPS (sat/nav)	yes	no
Media player	Audio/video player	no
Games	downloadable	3
Talk time	14 hours	3 hours

A New smartphones, like the iPhone (left), have a greater range of functions than earlier mobile phones (right)

Control systems

Electronic and mechanical systems are often called **control systems**. This is because they 'control' what things do. It is possible to integrate simple control systems with resistant materials to produce products in school that 'do something'.

It is not always necessary to understand how these systems work, but you might find it useful to know what they do and how they can be used. Many of these systems can be bought ready to incorporate into projects, requiring little or no assembly.

Any control system can be described in a simple flowchart. These flowcharts, or block diagrams, make the control system easier to understand. All control systems have three major functions: input, process and output.

D A control systems diagram

- An **input** activates a system. Common inputs include electrical components that react to environmental changes such as temperature, sound or light. They also include human inputs, such as turning a handle or moving a lever to operate a machine.
- The **process** describes what the control system does in response to an input, or how the input is converted into a useful output.
- The **output** is what happens after the input has been processed. Common outputs include electrical components such as bulbs, Light-emitting diodes (LEDs) and buzzers, and mechanical components such as gears, cams and levers.

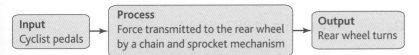

E A bicycle is an example of a mechanical control system

Integrating systems in school

In school, **electronic control systems** can be used to make products such as alarms and timers and used to control lights or LEDs.

Mechanical control systems are used to make things move, and are integrated into products where movement is required.

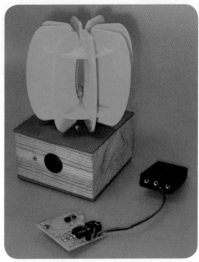

B A 'mood light', made using resistant materials and a colour changing LED. The circuit (shown) has a light sensor to turn the light on when it gets dark.

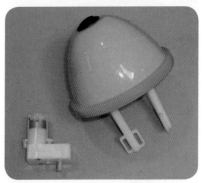

C A 'cocoa stirrer' made from vacuum-formed polythene, acrylic and an electric motor (shown). The motor turns the paddles, which then stirs the cocoa.

Activities

1 An electric bicycle, also known as an e-bike, is a bicycle with an integrated electric motor. Make a list of the advantages and disadvantages of an e-bike compared to a traditional bike.

2 Compare a modern-day car with one from the 1950s. Make a list of the features that are now standard on a modern car, but which would not have been available for the earlier vehicle.

Summary

- Products can be designed to have greater functionality.
- Control systems can be designed into resistant materials projects to make them more functional.

9.1

Case study: Famous designers

Objectives

- Be able to explain what is meant by the term 'design classic'.
- Become familiar with the work of famous designers.

Products are designed to provide solutions to real problems and to make life easier. Everything around you has been designed – from the chair you sit on, to the MP3 player you use to listen to music. Some products are totally new, but some are the result of **designers** improving existing products.

Some designers make products that work so successfully their designs become known as '**design classics**', and they become world famous.

Case study

Marcel Breuer (1902–81)

Marcel Breuer studied and taught at the Bauhaus (an art, design and architecture school) in Germany in the 1920s, where he pioneered the design of tubular steel furniture. Although his designs are nearly 100 years old, they still look **contemporary** and are used in homes and offices around the world.

One of his most widely-recognised designs is the bent tubular steel chair, later known as the Wassily Chair (Figure **A**), designed in 1925 and inspired by the curved tubular steel handlebars of his bicycle.

A The Wassily Chair, designed by Marcel Breuer

Case study

George Carwardine (1887–1947)

George Carwardine was an English engineer who designed the 'Anglepoise' desk lamp in the 1930s. The shade is attached to an arm that is jointed in two places. The design, which allows the light source to be easily adjusted, uses springs to keep the arm in place. Its design was widely copied by other manufacturers. The original design is still being manufactured today, including the Anglepoise Giant (Figure **B**), which is three times the size of the original!

B The Anglepoise Giant, designed by George Carwardine

Key terms

Designers: people who designs products.

Design classics: unique designs, which work perfectly and are often copied.

Contemporary: belonging to the present day.

Activities

1 Think about the products you use every day. Make a list, and choose three that you think could become future design classics. Explain your choice.

2 We are all aware of 'designer products'. Find out more about the designer of a product you use.

Case study

Arne Jacobsen (1902–71)

Arne Jacobsen was one of Denmark's most famous architects and designers. He enjoyed worldwide success with his simple, but effective, chair designs.

Following his design for the 'Ant' chair, in 1955, he designed the 'Series 7' (Figure C), which has become one of the most successful chair designs ever produced. He also designed the 'Egg' and 'Swan' chairs, which are considered 20th-century design classics.

C The Series 7 chair designed by Arne Jacobsen

Case study

Richard Sapper (1932–)

Richard Sapper is a German industrial designer who has designed products for a range of manufacturers, including the 'Melodic' kettle for Alessi and the 'Thinkpad' laptop computer for IBM.

In 1972, he designed the high-tech 'Tizio' work lamp (Figure D), which uses counter weights to balance the arms. The Tizio is featured in the Design Museum, London and the Museum of Modern Art (MoMA) in New York.

D The Tizio desk lamp designed by Richard Sapper

Case study

Philippe Starck (1949–)

Philippe Starck is one of the best-known contemporary designers in the world. He is an accomplished architect and an innovative product designer. He has been described as 'probably the most unusual, quirkiest and exciting designer of the past 20 years and likely to be for decades to come'.

He is probably best known for the 'Juicy Salif' lemon squeezer (Figure E), which he designed for Alessi in 1990.

E The 'Juicy Salif' designed by Philippe Starck

Activity

3 Prepare a presentation on your favourite designer, including examples of their designs.

Summary

● Everything we use is designed by someone.

● Some designers become famous because they produce unique designs that perform well.

When most of us think about designing, we think about products that satisfy our own needs and which fit in with our own lifestyles. However, everyone is different, and designers need to be aware of these differences when designing the products we use every day.

Exclusive design

Some products are designed to be **exclusive**. This means they are only designed for a specific group of people.

Children, the elderly and disabled people are all examples of user groups who have specific needs. Designers must be aware of these needs when designing products that target these groups.

Case study

StairSteady

'StairSteady' is the brainchild of Ruth Amos, and was originally her GCSE Resistant Materials project. The project came about as a response to the problem of a stroke victim needing to continue to exercise, but who was unable to use the stairs in his home. StairSteady is a handrail that moves up and down the stairs, providing support for the user. A device inside the handrail allows freedom of movement, but locks in place when weight is applied.

Ruth entered her prototype into the 'Young Engineer for Britain' competition in 2006. She won and was encouraged to develop and market her product. A local engineering company helped Ruth turn her GCSE project into a commercial product.

It was designed as a low-cost product to help people with limited movement, and is a much cheaper alternative to stair lifts.

A Ruth Amos with a prototype of 'StairSteady'

Inclusive design

The ideal product is one that everyone can use. **Inclusive** design means making a product suitable for the widest possible range of users, including all ages and abilities. Inclusive design is about making better mainstream solutions for everyone, rather than exclusive designs for groups with specific needs.

Case study

Worldbike

More than a billion people worldwide lack access to transport such as cars or trains. The bicycle is an ideal solution. However, many of the bikes used in developing countries are not designed for carrying large loads.

Worldbike designs and distributes low-cost bicycles, which are specially designed to carry heavy loads, cope with uneven surfaces and poor weather. These multi-use bikes are designed to be affordable and to be maintained and repaired locally. The 'longtail' or extended cargo rack can be used in a variety of ways, from carrying goods to market, transporting children to school or even being used as a taxi service!

B Big Boda: a multi-use 'cargo' bike designed for use in Kenya and Uganda

Case study

'Power to the people'

The Design Council, in association with the 'Helen Hamlyn Research Centre', has produced a range of case studies focusing on inclusive design. One study looks at the development of a range of new mainstream products for B&Q, to include the needs of older users. The challenge was to produce a range of DIY power tools that could be used successfully by both able-bodied users and those with limited physical movement. There were four solutions:

- A cordless screwdriver with a redesigned, more comfortable handle.
- A cordless drill with the heavy battery located away from the drilling unit.
- A redesigned jigsaw, which is more comfortable to hold, with better ergonomics.
- A small hand sander, which has been redesigned to fit the shape of the hand.

C A design for a cordless screwdriver with an easy-grip handle

Activities

1 a Choose one type of portable power tool available in your school workshop. Evaluate it from a user's point of view:
   - Is it easy to use?
   - Is it comfortable to hold?
   - Are there any features which make it difficult to use?
   b Redesign your chosen power tool and 'design-out' any poor features.

2 Consider your mobile phone. Create the profile of someone in the target group for this phone (include name, age, favourite food, job, type of person, where they live, etc.).

3 Design a mobile phone for the following user groups, and explain your design decisions:
   - young children aged 8–11
   - elderly people over the age of 65.

Link

Find out more information on Worldbike at: www.worldbike.org

Summary

- Some products are often designed to meet the needs of specific groups of people, for example children, the elderly and disabled.

- Some products are designed to meet the needs of everyone.

Case study: Product development

Objectives

- Understand how products develop over time to take advantage of advances in technology and materials, and to respond to the needs of the consumer.

Music players have become an important part of our lives. We have music players at home, in the car, and we even carry them with us when we go out. Music players have changed a lot since their introduction in the late 1800s, and it can be quite difficult to follow their development over the years because they now work very differently to the original devices.

Music players have changed because of **technological advances**, **new materials** and **automated production processes**. The development of newer products has also changed the way we use them. For example, the use of batteries, together with the development of plastic moulding technology and the miniaturisation of electronics, has meant that music players have become truly portable.

Case study

The gramophone

The wind-up gramophone (Figure **A**) revolutionised music technology. Music was recorded on to discs or 'records' made from shellac (an early form of plastic), but these were often very brittle. The music was recorded on to grooves on the surface of the record.

The record player

The widespread introduction of electricity saw the development of the electric record player (Figure **B**) in the 1950s. The technology was basically the same as the gramophone, but advances in materials saw the introduction of a synthetic plastic called polyvinyl chloride (PVC), which was now used to make the records. The grooves on 'vinyl' records could be much closer together, so that more music could be recorded on to the disc.

The tape cassette player

Cassettes, made from magnetic tape, became popular in the 1970s. They were much smaller than vinyl records and, protected in a plastic case, were far more durable. Because of the small size of the cassettes, tape cassette players (Figure **C**) also became much smaller. The introduction of reliable battery technology also meant that consumers did not need to rely on fixed electrical connections. In the late 1970s, the innovative 'Sony Walkman' was introduced, a compact version of the tape cassette player.

A An early gramophone

B A record player

C A tape cassette player

The CD player

CDs were introduced in 1981, designed originally to be the successor of the gramophone record for playing music. Their use for data storage came much later. CDs largely replaced the tape cassette, as they were far more durable and gave immediate access to different music tracks instead of the cumbersome method of fast forwarding and rewinding tapes. Portable versions were also introduced such as the 'Sony Discman' (Figure D).

D The Sony Discman

MP3, iPod and smartphone

Digital music technology has led to the development of a new range of music players. We can now download and store music on MP3 players, such as the fashionable iPod, which are small enough to be kept in a pocket. This has been helped by the miniaturisation of electronics and the development of new manufacturing processes. The demand for portable devices with greater functionality has also seen the rise of the smartphone, a small portable computer, which has the ability to play music files (Figure E).

Now, you no longer need a dedicated music player to play music!

E The iPhone, a pocket-sized computer that also plays music

Activities

1. In just over 100 years, the way we listen to recorded music has developed from wind-up gramophones to digital technology where music files can be accessed wirelessly from remote locations. Technology never stands still. Consider how we will listen to music in the future and design a 'future music player'.

2. The iPhone is a type of smartphone – a multi-function device.

   a. Explain the term 'multi-function device'.
   b. Make a list of other multi-function devices and explain what they do.

3. A lot of people still like to listen to music recorded on vinyl records. Find out why.

Key terms

Technological advances: newly-discovered ways of doing things.

New materials: newly-invented, or discovered, materials.

Automated production processes: computer-controlled manufacturing systems requiring very little human input (Figure D).

Summary

- Products can develop over time with the introduction of new technologies, new materials and new manufacturing processes. This is called 'technology push'.

- Products can also be developed in response to the needs of consumers. This is called 'market pull'.

Glossary

A

Adhesives: compounds used to chemically bond items.

Aesthetics: how something appeals to the five senses.

Alloy: a mixture of two or more metals.

Automated production processes: computer-controlled manufacturing systems requiring very little human input.

B

Batch production: making a quantity of a product before switching over to the next product.

Bending: forming an angle or curve in a single piece of material.

Blobject: a product designed using CAD or CAM to reduce designing constraints.

C

CAD CAM: a process used to produce products using computer-controlled output devices.

Casting: the process of making parts by pouring liquid metal into a mould and allowing it to solidify.

Ceramic: an inorganic material, normally an oxide, nitride or carbide of metal.

Composite: a material that is made from two or more material types that are not chemically joined.

Computer-aided design (CAD): the use of computer software to support the design of a product.

Computer-aided manufacture (CAM): using computers to operate machine tools.

Computer-integrated manufacture (CIM): using CAD to design a product, then CAM to manufacture it.

Constraint: something that limits what you can design and make.

Contemporary: belonging to the present day.

Continuous production: making the same product 24/7.

Control systems: a system with an input, process and output.

Corrosion resistance: the ability of a material to prevent damage due to chemical reactions with its environment.

Crating: using a box to provide guidelines for drawing.

Culture: how the beliefs, history and traditions have influenced a group within society.

D

Design brief: a short statement of what is required in a design.

Design classics: unique designs, which work perfectly and are often copied.

Design process: a sequence of activities carried out to develop a product.

Designers: people who designs products.

E

Electronic control systems: a system that uses electronic components.

Exclusive: designed for a specific target group.

Exploded view: a drawing that shows how the parts of a product fit together.

F

Features: details on the design.

Ferrous metal: a metal that contains iron.

Fixture: a device, similar to a jig, used to hold or position a work piece for machining that is attached to a machine.

Flow chart: a sequence of activities presented as a diagram.

Function: the task that a product is designed to do.

Functional testing: testing in real use, to check that the product does what it is meant to do.

Functionality: what a product does.

G

Grain: the direction or pattern of fibres found in wood.

H

Hardwood: a wood from a deciduous tree.

Hazards: things that could cause injury or harm.

I

Inclusive: meeting the needs of the widest possible range of users.

Injection moulding: the process of making plastic parts by forcing liquid plastic into a mould and allowing it to solidify.

Inorganic: a natural material, but one that does not grow.

Integrating systems: the use of technology to give a product greater functionality.

Isometric drawing: a 3-D drawing technique where horizontal lines are at 30° to the horizon.

J

Jig: a device used to hold or position a work piece for machining in order to achieve a consistent and repeatable end result.

K

Knock-down fittings: fittings used to join together flat-pack furniture.

L

Laminating: gluing thin strips of material together to make a thicker piece.

M

Manufactured board: a wood product made by processing or pulping wood particles or sheets.

Market: the group of potential customers who might buy the product.

Mass production: making the same product on an assembly line.

Mechanical control systems: a system that uses mechanisms to generate force and movement.

Model: a representation of a design idea.

Moral choices: decisions about what is good or bad.

Mould: a former used to shape a part.

N

Need: what the product you are designing must do.

New materials: newly-invented, or discovered, materials.

Non-ferrous metal: a metal that does not contain iron.

Non-renewable: something that is not replaced and will eventually run out.

O

Objective: based on facts, rather than opinions.

One-off production: making one of a product.

Orthographic drawing: a working drawing that shows the dimensions of the part.

P

Painting: applying a liquid that dries to form a coating on the part.

Patent: a legal protection for a design idea.

Personal protective equipment (PPE): equipment worn or used in the workshop to protect against hazards.

Photochromic: changes colour with changes in the level of light.

Piezoelectric material: a material that changes shape fractionally when a voltage is applied to it.

Plating: depositing a layer of metal using electrolysis.

Polishing: a physical process where the surface is rubbed or buffed to make it smoother.

Pollution: contamination of the environment.

Polymer: the correct name for what we normally call 'plastic'.

Primary research: finding out the information you need by yourself.

Product analysis: investigating the design of existing products.

Production line: a set of operations designed to produce a product as efficiently as possible.

Production plan: instructions on how to manufacture a product.

Properties: how the material performs in use.

Q

Questionnaire: a list of questions used to find out what lots of users want from the product.

R

Rapid prototyping: a CAD CAM system used to make accurate 3-D models.

Recycling: breaking or melting down the material so that it can be used in a new product.

Reduce: use fewer raw materials.

Refuse: do not accept designs that are not the best option for the environment.

Rendering: applying colour or texture to a sketch or drawing.

Repair: mending a product so that it lasts longer.

Research: collecting the information you need to be able to design the product.

Rethink: find other ways of designing the product to make it more environmentally friendly.

Reuse: using the product again.

S

Scale: the ratio of the size of the design in the drawing to the size of the finished item.

Secondary research: finding out the information you need by using data that someone else has put together.

Shading: creating different tones on a sketch or drawing.

Shape memory alloy (SMA): a metal that, after being bent out of shape, will return to its original shape when heated above its transition temperature.

Sketching: a quickly produced visual image of an idea.

Soft modelling: making a model using materials that are different to the final product.

Softwood: a wood from a coniferous tree.

Specification: a list of needs that the product must meet.

Standard: a document published by the BSI that lists all of the properties expected of a product and the tests that should be carried out.

Stereolithography (STL) file: a file format used for rapid prototyping and CAM.

Surface finishing: modifying the surface of the part in a useful way.

Sustainable manufacturing: products that are made using processes that are non-polluting and that conserve energy.

Sustainable materials: materials that are easily available and can be harvested, manufactured and replaced using very little energy.

T

Technological advances: newly-discovered ways of doing things.

Thermochromic: changes colour with temperature.

Thermoplastic: a polymer that can change its shape when heated.

Thermoset: a polymer that cannot change its shape when heated.

Threaded fastenings: mechanical parts such as screws, nuts and bolts.

Tone: lighter and darker versions of a colour.

V

Vacuum forming: forming a thermoplastic sheet over a mould, using heat and a vacuum.

Veneer: a thin layer of wood stuck on to a cheaper wood to improve its appearance.

W

Want: features that you would like the product to have.

Wasting: using tools to remove the unwanted material from a product.

Welding: a method where a joint is created by melting the contact areas of the parts being joined.

Index